INFORMATION TECHNOLOGIES

AND GLOBAL POLITICS

SUNY SERIES IN GLOBAL POLITICS

James N. Rosenau

editor

INFORMATION TECHNOLOGIES
AND GLOBAL POLITICS

*The Changing Scope
of Power and Governance*

edited by
JAMES N. ROSENAU
and
J. P. SINGH

State University of New York Press

Published by

STATE UNIVERSITY OF NEW YORK PRESS, ALBANY

For information, address
State University of New York Press,
90 State Street, Suite 700, Albany, NY 12207

Production and book design, Laurie Searl
Marketing, Michael Campochiaro

Library of Congress Cataloging-in-Publication Data

Information technologies and global politics : the changing scope of power and governance / edited by James N. Roseanau and J.P. Singh.
        p. cm. — (SUNY series in global politics)
    Includes bibliographical references and index.
    ISBN 0-7914-5203-4 (alk. paper) — ISBN 0-7914-5204-2 (pbk. : alk. paper)
    1. Power (Social sciences) 2. Information technology—Political aspects.   3. Information society—Political aspects. 4. State, The.   5. International relations.   I. Rosenau, James N.
II. Singh, J. P., 1961–   III. Series.

JC330 .I54 2002
327.1—dc21   2001048344

10 9 8 7 6 5 4 3 2 1

CONTENTS

FIGURES AND TABLES

FIGURES

TABLES

PREFACE

The essays that follow are the product of lengthy processes of conferring, writing, and rewriting on the part of their authors. From the beginning our goal was to bring together scholars who had contributed seminal work on the complex relationships between information technologies and global politics and who would then pool their expertise to clarify and further extend understanding of phenomena that seem ever more relevant to the course of events. While readers will have to decide for themselves whether we have, collectively, achieved this goal, the editors are persuaded that the lengthy collaboration has been a fruitful and rewarding experience.

The collaboration went through three face-to-face iterations. The first occurred in August 1996, with a workshop composed of both comparativists and international relations (IR) specialists in San Francisco sponsored by the Science, Technology, and Environmental Politics, section of the American Political Science Association (APSA). The second involved a one-day conference on Information, Power and Globalization held on January 24, 1998, in Memphis and funded by the BellSouth Foundation and the University of Mississippi. On this occasion outlines of the various projects and the papers they might yield were thoroughly discussed. During the summer of 1998 the authors circulated first drafts of their chapters to others, with each writer assigned as a "referee" for at least two other chapters. The third iteration of our collaboration, another workshop sponsored by the Science, Technology, and Environmental Politics section of the APSA, then occurred on September 2, 1998, at the Association's meeting in Boston. Subsequently, the final drafts were prepared and sent out for review by SUNY Press.

During this lengthy and complex process the editors incurred a number of debts which they are pleased to be able to acknowledge. Needless to say, we are grateful to APSA's Science, Technology, and Environmental Politics Section for sponsoring the two workshops on the issues and papers discussed in this volume. Woody Kay, Frank Laird and Vicki Golich were especially helpful with workshop logistics and funds. Likewise, without the support of BellSouth Foundation we could not have convened the second iteration under such conducive circumstances. Our thanks go also to the referees for SUNY Press who, while encouraging us with their approval, also provided feedback that helped to improve the book's quality. We also wish to express appreciation to several other scholars who were closely associated with the project either throughout or in particular workshops: Monique Moleon, Debora Spar, Tim Sinclair, and Virginia Walsh. Valuable help was provided by graduate assistants Ted Sirianni, Simona Folescu and Sarah Gilchrist. Finally, but no less gratefully, we wish to acknowledge Zina Lawrence, Priscilla Ross, Michael Rinella, and Laurie Searl at SUNY Press for their support and guidance through the many stages that have culminated in publication. Most importantly, we feel a deep indebtedness to our fellow authors for their participation and their readiness to rework their chapters in response to the many suggestions that came their way.

In addition, having worked together over such a long stretch of time, each editor wishes to express an overall feeling about the experience. J. P. Singh considers this project to be one of his most constructive scholarly experiences. The volume would have been impossible without the level of collegiality and intellectual commitment that all the participants made. In particular, J. P. Singh is profoundly thankful to his co-editor for his support throughout the project. Co-editorship has been an inspiring experience for Singh and he is thankful to Rosenau for his practical and creative insights, many of which have helped to shape this volume's quality. The volume has made Singh a strong believer in group projects and edited volumes! Singh also thanks his partner Chuck Johnson for his continual encouragement and important help with many aspects of this project. He is also grateful to his new colleagues at Georgetown University, especially Linda Garcia and Diana Owen, for providing him with an invigorating professional environment.

Likewise, Rosenau wants to emphasize that although alphabetic order has been used to list the editors, the order in no way reflects the division of labor between them. The idea for the project, the conferences, and this book originated with Singh. He and Rosenau collaborated throughout the project, but Singh hosted and funded the conferences that launched and sustained the project and he then gave considerable feedback to the authors in the process of their writing a final draft.

Lastly we are impressed with the thought that this book will not be the final word on information technologies, power, or global politics. The central thrust of the book is that these are forces that will continue to sustain transformative dynamics in the future. Hopefully we have provided a map with which to roam across this ever-changing landscape. Reading the map is no easy task—the pace of change being what it is in the information field—but the map is there and we take pride in having brought it into fruition.

ACRONYMS

AHCIET	Asociacion Hispanoamericana de Centros de Investigacion y Estudio de Telecomunicaciones
AMD	Advanced Micro Devices, Inc.
APSA	American Political Science Association
ASATs	Anti-satellite Weapons
ASEAN	Association of South East Asian Nations
ASETA	Asociacion de Empresas de Telecomunicaciones of the Andean Subregional Pact
AWACS	Airborne Warning and Control System
BIOS	Basic Input Output System
CBC	Canadian Broadcasting Company
CCD	Camouflage, Concealment and Deception
CGIAR	Consultative Group for International Agricultural Research
CIA	Central Intelligence Agency
CITEL	Inter-American Telecommunications Commission
COMTELA	Comision de Telecomunicaciones de Centro America
CPNs	Cross National Production Networks
CRTC	Canadian Radio-television and Telecommunications Commission
DARPA	Defense Advanced Research Projects Agency
DBS	Direct Broadcasting Satellites
DoD	Department of Defense (U.S.)
DRAM	Dynamic Random Access Memory
ENGO	Environmental Non Governmental Organization
EOSAT	Earth Observation Satellites
ERTS	Earth Resources Technology Satellite
FAO	UN's Food and Agriculture Organization

FCC	Federal Communications Commission
FinCEN	American Financial Crimes Enforcement Network
GATS	General Agreement on Trade in Services
GATT	General Agreement on Trade and Tariffs
GBT	Group on Basic Telecommunications
GDP	Gross Domestic Product
GIS	Global Information System
GM	Genetically-modified
GMO	Genetically-modified Organism
GNS	Group on the Negotiation of Services
GODs	General Obligations and Disciplines (WTO)
GPS	Global Positioning System
GSD	Ground Spatial Dimension
GSO	Geostationary orbit
HNGO	Humanitarian Non Governmental Organization
IARU	International Amateur Radio Union
IBRD	International Bank for Reconstruction and Development
ICANN	Internet Corporation for Assigned Names and Numbers
IGY	International Geophysical Year
IMF	International Monetary Fund
INTELSAT	International Satellite Organization
IOs	International Organizations
IPE	International Political Economy
IPR	Intellectual Property Rights
IRU	International Radiotelegraph Union
ISDN	Integrated Services Digital Network
ISI	Import Substitution Industrialization
ISP	Internet Service Provider
ITU	International Telecommunications Union, formerly International Telegraph Union
J-STARS	Joint Surveillance and Target Attack System
LDCs	Less Developed Countries
LEO	Low Earth-Orbiting (satellite systems)

MEDEA	Measurements of Earth Data for Environmental Analysis
MFN	Most Favored Nation
MITI	Ministry of International Trade and Industry
MNC	Multi-National Corporation
MTN	Multi-lateral trade negotiations
NAFTA	North American Free Trade Agreement
NASA	National Aeronautics and Space Administration
NGBT	Negotiation Group on Basic Telecommunications
NGO	Non-Governmental Organization
NICs	Newly Industrializing Countries
NOAA	National Oceanic and Atmospheric Administration
NTIA	National Telecommunications and Information Administration
OAS	Organization of American States
OECD	Organization for Economic Cooperation and Development
PC	Personal Computer
PCCIP	President's Commission on Critical Infrastructure Protection
PGRFA	Plant Genetic Resources for Food and Agriculture
PTC	Pacific Telecommunication Council
PTT	Post Telegraph and Telephone
PVOs	Private Voluntary Organizations, another term for NGOs
R&D	Research and Development
S&T	Science and Technology
SALT	Strategic Arms Limitation Talks
SAR	Synthetic Aperture Radar
SIGNIT	Signals Intelligence Surveillance
SPIN-2	Space Information—2 Meter
SPOT	Systeme Probatoire d'observation de la Terre
TCP/IP	Transmission Control Protocol/Internet Protocol
TIROS	Television Infrared Observation Satellite
TRIPs	Agreement on Trade-Related Aspects of Intellectual Property Rights
UNCTAD	United Nations Conference on Trade and Development
UNESCO	United Nations Education, Science and Cultural Organization

USTR   United States Trade Representative
VERTIC   Verification Technology and Information Center
VPN   Virtual private network
WRI   World Resources Institute
WTO   World Trade Organization

Important Terms

ANALOG: Based on a system in which numbers are represented by continuously variable and measurable physical variables, such as electrical signals.

BROADBAND: Comprised of or related to a wide band of electromagnetic frequencies, especially in communications.

DIGITAL: Of, relating to or based on calculations and logical operations with quantities represented as digits, commonly in the binary number system.

ENCRYPTION: To scramble (or encode) access codes to (computerized information) in order to prevent unauthorized access.

HYPERMEDIA: A computer-based information-retrieval system that allows a user to gain or provide access to texts, audio and video recordings, photographs, and computer graphics related to a particular subject.

INFORMATION: A collection of data or facts in usable form. (As opposed to an isolated and/or unrelated collection of data.)

INFORMATION TECHNOLOGIES: The technologies used in gathering, manipulating, classifying, storing, and retrieving data. This includes (but is not limited to): computers, telecommunication systems, broadcasting mediums, multi-media convergence, etc.

INFORMATION INFRASTRUCTURE: The basic facilities, services, and installations needed for the communication of information.

INTRANET: System in which information is stored on a secure computer that only serves other computers with designated access. The data is often displayed in the form of HTML (Internet) pages.

KEY: A table, gloss, or cipher that allows the user to decode or interpret data.

NETWORK: A communications system that shares information via telephone wires, direct cables, satellite links or other means.

NETWORKING: The exchange of information or assistance among members of groups or institutions, in both formal and informal settings.

PACKET SWITCHING: One method of data transmission in which small blocks of data are transmitted rapidly over a channel dedicated to the connection only for the duration of the packet's transmission.

TCP/IP: Transmission Control Protocol/ Internet Protocol (the common interconnection standard)

CHAPTER ONE

INTRODUCTION:
INFORMATION TECHNOLOGIES
AND THE CHANGING SCOPE OF
GLOBAL POWER AND GOVERNANCE

J. P. SINGH

With the steamship, the electric telegraph, the newspaper,
    the wholesale engines of war,
With these and the world-spreading factories he interlinks all
    geography, all lands;
What whispers are these O lands, running ahead of you,
    passing under the seas?
Are all nations communing? Is there but going to be one
    heart of the globe?
Is humanity forming en-masse? For lo, tyrants tremble,
    crowns grow dim,
The earth, restive, confronts a new era, perhaps a general
    divine war,
No one knows what will happen next, such portents fill the
    days and night;

—Walt Whitman, *Songs of Parting*

It is somewhat ironic, and a tad unpoetic, to note that in 1865 as Walt Whitman conjectured about the whispers passing under the seas, across the Atlantic in Paris the industrializing Western powers met to found the International Telegraph Union.[1] Zacher echoes Whitman's concerns in pointing out later in this volume that capitalism in general, the world-spreading factories, and the electric telegraph in particular, came with a "mandate for interconnection." The workings of that mandate have spiraled outward for 135 years.

1

This volume studies the relationship between information technologies and global politics, a relationship whose intricacies and the broad meanderings are still being understood. For this reason, the authors return to the bricks and mortar of political science: power and governance. Reflecting a view offered by many international relations researchers, they grapple with the way the spread of information technologies is shifting power and the locus of authority away from the state. The intent and contribution of the book are to show how this conclusion is valid with respect to information technologies by examining several issue-areas. The overall context of the volume is, of course, what information technologies have now wrought—global information networks. Networking, entailing communication and information exchange, is changing both the way power is exercised and governance is organized in global politics.

Information technologies in this volume refer to all technologies that help to produce, gather, distribute, consume, and store information. These may include, though are not limited to, print and broadcast media, telecommunications (telephone, fax, Internet, World Wide Web, etc.), channels of communication (satellite, different types of cable including fiber optics), computers, and storage devices (DVD, CD-ROM). Except for Aronson (chapter 2) and Rosenau (chapter 11), the authors focus on specific information technologies and issue-areas.

This chapter puts the rest of this volume in a theoretical and, where necessary, a historical, perspective. The chapter shows how "multiperspectival" identities, actors, and issues are supplementing national identities, states, and the salience of high-politics, or strategic security issues, in global politics. The chapter first discusses how the rise of information networks is facilitated by changes in technology. The chapter then turns to the changing scope of power. It shows how power with respect to technology needs to be understood as much in terms of capability (instrumental and structural power) as in terms of the ability of information technologies to constitute new identities and agendas—what this chapter terms "meta-power." Finally, this chapter discusses the scope of governance. The argument is that the locus of authority, order, and legitimacy are shifting away from the state toward pluralism and actor advocacy.

The volume's authors are attempting to grasp the quotidian and the transformational effects of information networks over global politics. Information technologies and global politics have been studied before, and there are a few classics in the field (Zacher with Sutton 1996; Sandholtz 1992; Krasner 1991; Cowhey 1990; Aronson and Cowhey 1988; Gilpin 1981). However, debates continue on basic questions such as the impact of the information networks on the identities of actors, and what these actors do unto each other in areas such as power, authority, and governance.[2] The authors in this volume try to build upon the nascent body of literature that endeavors to grasp these effects (for example, Keohane and Nye 1998; Deibert 1997; Der Derian 1990; Luke 1989). Fundamental changes, as those being brought about by information technologies, take time and are thus hard to analyze when they have just begun. By critically analyzing issues of power and governance, and by building on conclusions offered by international relations scholarship, it is hoped that a few in-depth answers may be provided on the relationship between information technologies and global politics.

THE RISE OF NETWORKS

The import of transactions conducted over networks is such that all conceptual frameworks now speak of networked organizations. In one form or another, scholars of various hues refer to these networks in speaking of the actors that international relations scholars study. Rosecrance (1996) refers to the virtual state; Deibert (1997) and Arquila and Ronfeld (1997) to networked security; Spar with Bussgang (1996) to networked marketplaces; Gereffi (1995) to networked transnational enterprises; Mathews (1997) and Keck and Sikkink (1999) to NGO-based advocacy networks; and all forms of networked organizations as preeminent in world political economy are referred to by Aronson (chapter 2), Keohane and Nye (1998), and Castells (1998, 1997, 1996). In Castells's (1996, 469) words: "Networks constitute the new social morphology of our societies, and the diffusion of networking logic substantially modifies the operation and outcomes in processes of production, experience, power and culture."

Technologies propose change; they do not determine it. The effects of networking on states, businesses, and international organizations transcend any kind of technologically deterministic logic. Nonetheless, two developments are important for understanding how technologies proposed the rise of networks that replaced earlier organizational forms. These developments are: digitization and the fall in marginal costs. Skeptics of the effects of information technology often question, not just the effects, but also the technological changes that facilitate them. It is thus important to understand technology as well as its effects on power and governance in the context of this volume.

Digital technology changed the way information industries were organized.[3] Historically, different types of information technologies evolved as distinct industry types dominated by one or more firms. The vertical dimension of Figure 1.1a captures the tasks performed by the different types of information industries. Vertically integrated industries developed different pipelines for different functions needed to deliver information. (Aronson discusses similar processes by referring to conduits in the next chapter.) Thus, the telephony industry deployed a combination of transmission media with high bandwidths (to carry messages over long distances) and narrow bandwidth copper wires (to deliver voice messages to particular homes, known as the "local loop" in industry jargon). The inability of

FIGURE 1.1
THE INFORMATION INDUSTRY

1.1a
Influence of Analog Technology

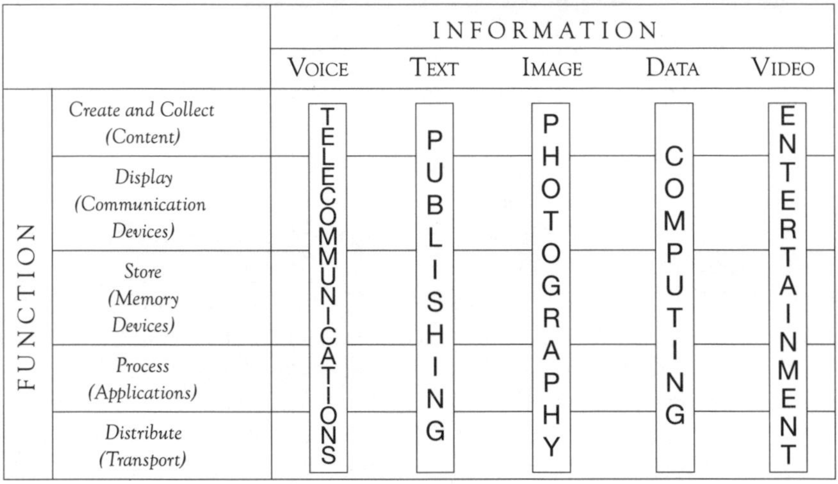

		INFORMATION				
		VOICE	TEXT	IMAGE	DATA	VIDEO
FUNCTION	Create and Collect (Content)	T E L E C O M M U N I C A T I O N S	P U B L I S H I N G	P H O T O G R A P H Y	C O M P U T I N G	E N T E R T A I N M E N T
	Display (Communication Devices)					
	Store (Memory Devices)					
	Process (Applications)					
	Distribute (Transport)					

1.1b
Influence of Digital Technology

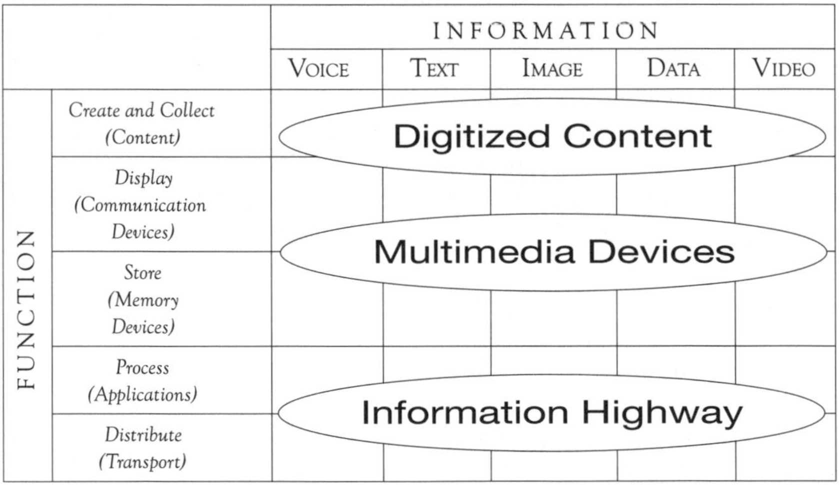

		INFORMATION				
		VOICE	TEXT	IMAGE	DATA	VIDEO
FUNCTION	Create and Collect (Content)	Digitized Content				
	Display (Communication Devices)					
	Store (Memory Devices)	Multimedia Devices				
	Process (Applications)					
	Distribute (Transport)	Information Highway				

Source: Sheth and Singh (1994)

these networks to carry high amounts of data (the first parallel horizontal frontier that the telephony industry broached) over the local loop is still being felt. In a few national markets, telecommunications providers were specifically barred from providing any services other than telephony in return for monopoly privileges. Similarly, cable television later distributed its content on a network capable of handling high bandwidths, but not particularly capable of switching it (as in telephony), because of the separation of industry types. Analog technology thus helped to separate voice, text, image, data and video industries.

Digital technology has undone the technological logic behind separate industry types and pipelines. This in turn has also spurred multimedia interactive instruments and fiber-optic cables capable of carrying all types of messages at high speeds and low costs (including over the local loop). Earlier technology was "analogous" (therefore the term analog) to sending information in electrical waves and was time-consuming and often inefficient. New technology allows information to be encoded in streams of binary digits (digitization) which can be sent efficiently and at relatively low cost over long distances. Digitization impacts all aspects of the information industry allowing various types of media (voice, text, image, data, and video) to be digitized and sent over the same pipeline and accessed by a single instrument. As shown in figure 1.1b, this offers the potential for horizontal integration of industry types. Even though the telecommunications industry is still catching up with this horizontal integration, multimedia interactive devices are already a trillion-dollar industry.

The vertical and horizontal integration of pipelines due to digitization is expanding and deepening information networks. The expansion is coming as different types of vertical pipelines merge. For example, the fact that cable networks can now accommodate telephony and vice versa allows for better and expanded geographic coverage. Deepening occurs due to horizontal integration, allowing for a variety of functions to be performed over the same network with the use of a multimedia device. It is this vertical and horizontal integration, whose genesis lies in digitization, that is leading to the oft-discussed information superhighway. Negroponte (1995, 231), in a popular book written about digitization, notes that the "information superhighway may be mostly hype today, but it is an understatement about tomorrow."[4] This raises the stakes for the firms involved in physically laying out the information superhighway, and also underscores the importance of the number of transactions conducted over this infrastructure. Kim and Hart (chapter 6) reflect the former concern in noting the battle over intellectual property rights and Aronson (chapter 2) explains the transactions in terms of what he labels conduit and content issues. The chapters by Zacher, McDowell, and Singh deal with similar themes of market access, user demands, and telecommunication providers' rivalries.

The second technological feature of importance is the way technological innovation pushes down the unit cost of products. Anyone who buys a computer one year to see its price halved the next season is familiar with this logic. Digital technology comes with high fixed costs and minuscule marginal costs, a development popularly known as Moore's Law (after Intel Chairman Gordon Moore). For example, a computer disk, once produced, can be reproduced a million times over at negligible cost. Selling one disk for $1 million is hard, but a million can sell for $1 each. Success for information age products derives in large measure from the ability to rapidly generate large volumes of demand in a short time. Microsoft Windows 95, with $700 million of sales on its first day, is one dramatic example. It helps to explain the push by firms like Microsoft to develop global standards and intellectual property rights in their favor. Superior and better microprocessing chips are also helping to do increasingly complicated tasks at faster and cheaper rates.[5] Generating large amounts of demand helps to recover costs.[6]

Declining marginal costs are facilitating network deepening and expansion at a rapid rate. A poor country with access to some capital can, if it has the political will, leapfrog the technological frontier by using inexpensive satellite based terminals and bringing a variety of multimedia services to remote areas. Furthermore, network distance matters less and less. Consumers in the United States are familiar with this logic through the one-rate long distance plans which replaced distance sensitive plans of the past. But the extent to which networking comes about, and the global impact it has, cannot be measured by technological developments alone. The fundamental points made in this volume about the impact of information technologies are thus rooted in the political, economic, and cultural context of their deployment and analyzed through the lens of the changing scope of power and governance.

THE CHANGING SCOPE OF POWER

Global politics are inherently relational. Equations of power can be simplified to "who does what to whom." This may entail who is empowered versus disempowered (instrumental power); who is constrained in a given situation versus who gets to write the rules (structural power); and, finally, how basic identities, interests, and issues themselves are reconstituted or transformed in particular historical contexts, in turn redefining other relations of power (called meta-power here). To political science's traditional notion of instrumental or structural power, this volume adds the notion of meta-power to which, one way or another, this volume's authors allude. This section explores how these three types of power may be understood in relation to information technologies. Two arguments are extended: the way information technologies are enabling formerly underprivileged groups to play a role in global politics, and the way in which all actors' identities and issue-areas are being reconstituted.

INSTRUMENTAL POWER

Instrumental power focuses on the capacity or capability of power holders to effect particular outcomes. Information technologies, or any technology for that matter, are then forces that enhance these capabilities.[7] This was one of the first ways in which political scientists and policy makers examined the relationship between information technologies and power. Information technology enhances the capabilities of traditional global actors, like states and firms, but it also empowers other actors (like transnational social movements or terrorist groups) and may even offer a few surprising insights into who is getting empowered and disempowered in global politics.

Early conceptualizations of the impact of technology on power, in scholarship and public policy, revolved around notions of instrumental power.[8] The instrumentality of the telecommunication infrastructure is apparent historically in the U.S. Department of Justice's concerns about AT&T's use of monopoly power in the early 1940s. These concerns were initially sidetracked, as the infrastructure was deemed too important for national security to warrant an investigation. The post-war investigation led to restrictions on AT&T in 1956 to stay out of information services. The Pentagon, however, continued to support AT&T until its breakup, deeming the latter a threat to national security.[9]

Security concerns began to spill over into the economic realm in the 1960s when enhancing national wealth through information infrastructures surfaced and became linked to concerns about national power. By the late 1960s, the Japanese had encouraged an entire school of scholars to think about "johaka shakai" or information society (Snow 1988). Powerful ministries such as MITI (Ministry of International Trade and Industry) got involved as the economic implications of information technologies became important (Aronson and Cowhey 1988). The French noted explicitly by the late 1970s that unless they enhanced their information infrastructure, they would be left behind politically and economically. A report written for President Valéry Giscard d'Estaing began with the following statement: "If France does not respond effectively to the serious new challenges she faces, her internal tensions will deprive her of the ability to control her fate. The increasing computerization of society is a key issue in this crisis and could either worsen it or help solve it" (Nora and Minc 1980, 1). Studies from international organizations like the ITU and IBRD, beginning in the 1960s, also advocated that developing countries could accelerate their pace of economic growth by expanding their information infrastructures (Saunders et al 1994/1983; ITU 1984; Hudson et al. 1979).

Instrumental power concerns were most obvious in the 1980s in national debates about economic competitiveness. The case of France was mentioned earlier. However, the prioritization of the information infrastructure also took place at a

European-wide level with the European Commission's 1987 Green Paper and the 1993 White Paper laying down the necessity of having suitable information superhighways for European industry (Bruce et al. 1988; Fuchs 1993; Sandholtz 1992; Singh and Sheth 1997; Wellenius and Stern 1994, part IV). In Asia, countries like Japan, South Korea, Singapore, and India launched national initiatives to boost infrastructural development (Singh 2000 and 1999; McDowell 1997; Melody 1997; Petrazzini 1995; Larson 1995; Wellenius et al. 1994; Sisodia 1992; Bruce et al. 1988). South Korea's waiting list of nearly 700,000 for telephones in the late 1970s was reduced to provide universal or country-wide coverage by 1987. Singapore began to see an information infrastructure as vital to its entrepot role, articulated best through its "second industrial revolution" launched in the 1980s. Singapore's vision of an "intelligent island" is exemplified in the availability of 100 percent ISDN in 1989 and expected availability of 100 percent fiber optic broadband network by 2005.

The competitiveness concerns in the United States reached a crescendo in the late 1980s with the growing fears about competitiveness in key sectors like automobiles and steel (Tyson 1992; Tyson and Zysman 1983; Hart 1992). These concerns spilled over into infrastructural development (Aronson 1992). Reports pointed out deficiencies in the U.S. infrastructure while pointing out others' strengths (see NTIA 1991; NTIA 1988; NTIA 1985).[10] Policymakers were soon implementing initiatives to stem the tide. Even laissez-faire minded Reagan boosted federal funding for Sematech in Texas to thwart the decline of competitiveness in the semiconductor industry. Vice President Gore years later touted schemes for a National Information Infrastructure, a broadband initiative. The Telecommunications Act of 1996, now considered a failure, was designed to help expand the information infrastructure.

Instrumental concerns about economic power and technologies were joined by traditional concerns like security and political change by the end of the century. Conceptions of security changed in two ways. First, information technologies were deployed to enhance capabilities in tasks ranging from making of "smart weapons" to organizational ones like defense preparedness (Arquila and Ronfeldt 1997; Deibert 1997; Nye and Owens 1996). Second, protecting national information infrastructres against varied threats became a regular concern of states. The latter ranged from individual hackers getting hold of crucial information to well-publicized cases of "cyberwars." It was discovered in 1998 that Russians, for example, got access to the Pentagon's computers lifting information, the extent of which even the U.S. government did not know. Instrumental power advocates focused on enhancing capabilities to protect these infrastructures. While recognizing that central control by states over decentralized networks is improbable, instrumental notions of another sort are apparent in the solutions. A White House report on ensuring insfrastructural reliability against natural and human calamities noted, "The national interest can only be served with the sustained engagement of industry, utilities, the public, and government at all levels" (Executive Office of the President 1997, 3). How this other type of central coordination can be effected beyond acts of moral suasion is not apparent.[11] Deibert shows in chapter 5 that conjectures about trying to enhance national power are unlikely to be sustainable in an age of networks.

Political change was also influenced by information networks. Unlike the 1980s, when the United States was seen as lagging behind in infrastructural provision, confidence re-emerged about the country's political role in the next decade. The United States got a renewed lease on its hegemony after the fall of the Soviet Union, East Asian financial crises, and Europe's economic slowdown. "In a world in which the meaning of containment, the nuclear umbrella, and conventional deterrence have changed, the information advantage can strengthen the intellectual link between U.S. foreign policy and military power and offer new ways of maintaining leadership in alliances and ad hoc coalitions" (Nye and Owens 1996, 20). The country had come round a full circle, from being an infrastructural laggard to possessing an information advantage.

The instrumental features of information technologies, of course, extend beyond state concerns. The way that these technologies empower less privileged groups is especially important in recognizing the promise of technology in instrumental contexts.[12] The spread of democracy in Russia, as Rosenau points out in chapter 11, was in crucial ways tied to the proliferation of information networks and accessibility of information for individuals and groups. While Litfin (chapter 3) and Braman (chapter 4) go beyond merely positing instrumental contexts, both of them do acknowledge how technology may empower NGOs (Litfin) and result in defense technologies enabling civil groups (Litfin, Braman) or even terrorist groups (Braman).

Chapter 10 offers a counterintuitive result flowing from instrumental notions of technology and underprivileged groups. Contrary to current wisdom, developing countries came away with significant concessions from developed countries during the recent WTO telecommunications negotiations. This results from the presence of multiple issues and actors in the global economy, themselves a result of the information age, providing more alternatives to weak states instead of the 'take it or leave it' scenarios that usually confronted them in the past.

Spar (1999) elsewhere has argued that MNCs, especially those producing consumer goods, have an incentive to improve their human rights practices as a result of what she calls the "spotlight phenomena," or the increasing tendency of information networks to spread the word about human rights abuses quickly. The spotlight phenomena that Spar mentions is related to the proliferation of "public eyes" that Litfin outlines in chapter 4.

Information technologies, in particular, are making us re-examine and rein-force several cherished ideas about instrumental power. First, we need to go be-yond a focus on states and firms. State capabilities are no longer dependent, for example, on merely using these technologies, but also from working in concert with a host of actors in enhancing their power. Second, the ways in which non-state actors are privileged is important. There is ground for optimism as human rights practices improve, democracy spreads, and the underprivileged make gains. However, the latter argument must not be overstated. States and firms have better access to information technology and information than others. Hackers, terrorist groups and nations engaging in acts of information warfare are also difficult to control. That instrumental power can have negative as well as positive conse-quences unhinges the original positive connotation of instrumental power.

STRUCTURAL POWER

Instrumental and structural power both deal with capabilities, but whereas the former emphasizes the ability to effect outcomes, structural power is about the ability to effect rules and institutions that govern these outcomes. The famous formulations of structural power in international relations include Waltz's (1979) positing of nation-states as a structural hierarchy of power capabilities, exhorta-tion by Keohane and Nye (1977) examining structural power within issue-areas at the macro-level of the world system, and Cox's (1987) contribution to how mate-rial capabilities, ideas and institutions constrain human action. By definition, structural power is concerned with the constraints and the fit of particular activi-ties with given institutions, or the ability to change the institutions rather than with notions of empowerment.

Structural power issues, like their instrumental counterparts, used to be about states and firms. In many ways, they continue to be so. But information technologies are making us appreciate the ways in which information, knowledge and ideas shape these structures and, in turn, human behavior. Strange's (1991, 1988) four structures (security, production, finance, knowledge) constraining op-tions for international actors are determined by states, markets and technology. Cox borrows from Gramscian thought to note that material and institutional structures cannot be examined without reference to ideational hegemonic con-texts.[13] Finally, Rosenau (1997) shows that individuals are now performing in-creasingly skillful tasks amidst complicated issue structures, in part sustained by information technologies.

The reciprocal relationship between technology and structures is noted in three ways. First, technology influences the structures of security or economic af-fairs. Second, existing structures or institutions shape technologies themselves.

An in-between case may be the so-called best fit scenarios between particular technologies and governance institutions.

The case of technology shaping structures is made foremost in radical scholarship.[14] Relatively doctrinaire Marxian schema posit so-called 'forces of production' (including technology) to be essential in the unfolding of history, shaping social relations (as between capitalists and workers). The dialectical relationship is held in place by the superstructure, including the state that "exists to guarantee the reproduction of these social (including economic) relations as a whole" (Fine and Harris 1979, 95). Following Marxian footsteps, Winner (1977, 82) concludes that "technologies are structures whose conditions demand the restructuring of their environments." To Winner, these structures and processes represent a "technological order" where technological adaptation is a "reverse adaptation," co-opting individuals into its workings. Unlike Faustian instrumental versions, Winner presents a technological Frankenstein—technology out of control of human agency (Singh 1994). Ends no longer follow from the means. "The true price is loss of freedom," a major theme that gets reiterated by historians such as Polanyi (1944) Hobsbaum (1968) and scholars like Postman (1985) and Castells (1996, 1997, 1998). In a famous formulation by Polanyi (running contrary to Zacher's and Rosenau's in this volume), technology does not create freedoms, the so-called freedoms serve the purposes of technology owners. "There was nothing natural about laissez-faire; free markets could never have come into being merely by allowing things to take their course" (Polanyi 1944, 139).[15] The radical tradition is reflected by Comor in this volume who argues that, in the context of capitalist social-economic relations, information technologies are being used to 'deepen' and 'broaden' the commodification of our daily lives. Capitalism here directly affects media contexts, sustaining the market system, generating a range of tensions and potential contradictions.

In a "technological order," information networks are governed not by an invisible hand, but by an invisible master. Network interconnections, countermovements, and interdependencies lead to a hierarchical positing of structural power with limited choice for human agency. A slightly different notion of structural power comes from those who see existing structures constraining the use of information technology. Structure determines what technology can or cannot do, instead of vice versa. Rosenau emphasizes this when he notes in chapter 11 that technology is neutral but that its use is shaped by the environment in which it finds itself.

The property rights literature in general has examined how rules or rights governing property lead to different uses of technology. Why is it that England took the lead in deploying technology that originated on the continent? North and Thomas (1973) argue that it was because of the nature of England's property rights that fostered industry.[16] When these property rights are not captured by small influential groups, the benefits are the greatest (Olson 1982, 1965). There are thus contexts in which technology and structures, or political-economic institutions, adapt to each other. Again, information technology can claim older counterparts. Landes's (1969) famous formulation of industrial technology and economic growth, *The Unbound Prometheus*, was about how English institutions and culture fostered constant innovation and technology usage while continental Europe lagged behind. Since then, there have been several best fit arguments noting how particular technologies falter or are adapted because of the institutional mix in place (Hart 1992; Best 1990; Sabel 1982).

Comparative analysts examine how and why it is easier for a few countries to expand their information infrastructures while others lag behind. Types of states and other institutions are examined to posit levels of infrastructural provision (Singh 1999; Levy and Spiller 1996; Wellenius et al. 1994; Duch 1991). Kim and Hart (chapter 6) cite similar literature to show how the institutional mix in the United States might be best suited to take advantage of Wintelist information technologies, epitomizing in the synthesis existing between Microsoft Windows and Intel.

Analysts also note that information networks are decentralized organizations ill suited for institutional contexts that try to centralize or control information flows. Frequent media accounts abound about Singapore or China controlling information flows. There are also subtler variations. Daniel Bell (1980) cited Stanley Hoffman's term *societe bloquee* or "a society that has become increasingly rigidified in its bureaucratic and political institutions" for characterizing France as it readied itself for information networks in the late 1970s. Two decades later, *The New York Times* (February 11, 1997, A1) characterized the country's dilemma as follows: "In other words, how do you leap into the age of the Internet and still remain French?" Deibert (chapter 5) reflects these concerns about the decentralized and nonhierarchical nature of information networks to show how these networks will themselves foster particular institutions.

In summary, three notions of structural power have been noted—one where technologies shape institutions, one where institutions determine technological use, and lastly the best fit scenarios where institutions and technology shape each other. In each case, information technologies may not increase the structural power of traditionally powerful actors.

META-POWER

Technologies not only impact existing actors and issues but, as an increasing body of knowledge notes, networked interaction itself constitutes actors and issues in global politics. If we merely focus on actor capabilities and take their identities and interests as given, as most instrumental and structural power versions do, the transformation being brought about by information networks is missed. Networking is highly interactive. Meta-power thus refers to how networks reconfigure, constitute, or reconstitute identities, interests, and institutions. Such power is referenced in this volume by Braman (chapter 4) in drawing attention to meta-technologies and genetic power; Litfin (chapter 3) to constitutive power; Kim and Hart (chapter 6) to meta-power and post-structural power; and Deibert (chapter 5) in referring to the constitution of 'collective images' about security. These authors also note that as ideas, interests and institutions are reconstituted, power shifts away from the original powerholders. The very nature of power itself and the actors who wield it is also changed.

The distinction between meta-power and instrumental or structural power made earlier is now increasingly recognized by those working within and outside traditional international relations scholarship. Interestingly enough, even neo-realists implicitly recognized the notion of meta-power early on. Gilpin (1981, 39), for example, distinguishes between regular interstate interactions and changes *in* systemic governance versus fundamental changes *of* the system dealing with "the nature of the actors or diverse entities that compose an international system." He notes that the latter change is understudied but that it is "particularly relevant in the present era, in which new types of transnational and international actors are regarded as taking roles that supplant the traditional dominant role of the nation-state, and the nation-state itself is held to be an increasingly anachronistic institution" (Gilpin 1981, 41). However, while recognizing these transformations, Gilpin does not deviate much from the instrumental notions of power.

Krasner (1985) refers directly to meta-power when noting post-colonial Third World advocacy. Meta-power would allow these states to steer the structure and rules of the market-based liberal international economy toward an authoritatively distributive structure. Krasner sees Third World calls for the creation of UNCTAD, New International Economic Order, and New World/Information Communication Order as strategies for power maximization. He then returns to a familiar conclusion—meta-power itself depends on capabilities, the Third World must suffer what it must. It can not reconstitute the system.

A few neoliberals, too, come close to delineating a notion of meta-power. Keohane and Nye (1988) point out the ascendance of soft power, or power through persuasion and attraction rather than force, as a new salient feature of global politics when information networks proliferate. The cognitive and interpretative insights offered by other neoliberal scholars also address issues of interest and preference formations (Haas 1989; Sell 1998; Odell 2000).

Nonetheless, most neoliberal and neorealist analysts, with few exceptions, take their cues from rational choice analyses, in which the identities and interests of actors, mostly nation-states, are posed ex-ante. Gilpin's concern is not how identity gets constituted but how new types of actors (be they empires, nation-states, transnational enterprises) influence the international system. Krasner's meta-power is about weak nation-states clamoring for power in the world system. Keohane and Nye's soft power is related to actor interests that have been taken as given. These static notions are under scrutiny by analysts situating their arguments in historical sociology, a growing tradition in international relations, now called "the constructivist turn."[17] The challenge is best summarized by one of constructivism's chief proponents, Alexander Wendt (1992, 393–394): "Despite important differences, cognitivists, poststructuralists, standpoint and postmodern feminists, rule theorists, and structurationists share a concern with the basic sociological issue bracketed by rationalists—namely, the issue of identity- and interest-formation. . . . They share a cognitive, intersubjective conception of process in which identities and interests are endogenous to interaction, rather than a rationalist-behavioral one in which they are exogenous." Wendt recognizes that there are scholars, especially in the neoliberal tradition, who have craved such analysis, and he is answering the critics of constructivism as well as trying to bring about a gestalt shift in them. Keohane (1988), years earlier, had called these traditions reflectivist. While appreciating the historical contextuality of intersubjective interest and identity formation, Keohane (1988, 381) noted that "the sociological approach has recently been in some disarray, at least in international relations: its adherents have neither the coherence nor the self-confidence of the rationalists."

Keohane's critique notwithstanding, other disciplines have long offered the kind of empirical insights that he demands. Halbwach's (1992/1941) early work on collective memory showed how images and symbols that societal groups hold can be traced historically and shape the preferences of group members.[18] Halbwach (1992/1941, 189) concludes that "all social thought is essentially a memory and that its entire content consists only of collective recollections or remembrances. But it also follows that, among them, only those recollections subsist that in every period society, working within its present-day frameworks, can reconstruct." Berger and Luckmann (1966) call attention to primary and secondary socializations to argue that reality is a social construction.[19] "Identity is formed by social processes. Once crystalized, it is maintained, modified, or even reshaped by social relations" (Berger and Luckmann 1966, 173). Anthropologist Geertz (1973, 20) was a forceful early advocate: "To set forth symmetrical crystals of significance, purified of the material complexity in which they are located, and then attribute their existence to autogenous principles of order, universal properties of the human mind, or vast a priori *weltenschauungen*, is to pretend a science that does not exist and imagine a reality that cannot be found." Putting it bluntly, "there is no such thing as a human nature independent of culture" (Geertz 1973, 49). Sociologist Castells (1997, 7) would agree: "It is easy to agree on the fact that, from the sociological perspective, all identities are constructed. The real issue is how, from what, by whom, and for what."

While postmodernists deliberately eschew what they term "instrumental empiricism," they provide a conceptual antidote to Keohane's universal rationalistsic notions. Foucault's analyses (1977, 1970) painstakingly reconstruct the social circumstances that privilege particular knowledge. All forms of knowledge then reveal micro-power relations carrying subtle means of co-opting or marginalizing individuals. Said (1978, 40–41), acknowledging an intellectual debt to Foucault, shows how colonizing Europe in fact created the Orient as a location, idea, and homogenous culture: "Knowledge of the Orient, because generated out of strength, in a sense *creates* the Orient, the Oriental, and his world. . . . Orientalism, then, is knowledge of the Orient that places things Oriental in class, court, prison, or manual for scrutiny, study, judgment, discipline, or governing." The construction and domination of the Orient are inextricably linked.

Indeed, while the constructivist turn is somewhat new in international relations scholarship, conceptually it stands to benefit from constructivist claims made elsewhere. To refine the concept of meta-power, this is a valuable exercise. The constitution of identities and interests in global politics may be related to similar conceptualizations by other social theorists.

The link between information networks and constructivism can now be made explicit. The collective meanings that actors hold about themselves, or meanings imposed upon them, are shaped by networks and in turn influence networks. But the constitution and effects of such identity formation remain contested among scholars. A few theorists see technology as merely playing a catalytic role in accelerating or reinforcing extant or incipient processes. Others see technologies as allowing for new types of identity and collective meanings. A quote from Said (1978, 26) is illustrative: "One aspect of the electronic, postmodern world is that there has been a reinforcement of the stereotypes by which the Orient is viewed. Television, the films, and all the media's resources have forced information into more and more standardized molds." Here technology remains neutral, reinforcing existing stereotypes.

Litfin (chapter 3) offers a nuanced empirical case of the complicated, and somewhat serendipitous, processes governing network effects. Building on Foucault and on Jeremy Bentham's ideas of the Panoptican, where a "disciplinary gaze" monitors and conditions the human behavior, Litfin notes that the diffusion of networks leads also to the decentralization of this gaze and the proliferation of "public eyes." In understanding such shifts, therefore, we must move beyond analyses which view technology only in an instrumental fashion. Litfin shows that information networks are in fact facilitating a new social episteme that not only changes the definition of issues in question (security, environment and human rights in her chapter) but also allows for new actors (NGOs in her case) to start playing key roles in global politics.[20] Her analysis, therefore, illustrates "both of the ways in which technological change can alter international reality: instrumentally and constitutively."

Litfin's makes us question the technological neutrality assumption where technology merely facilitates preexisting actors and issues and does not propose new identities or action. This, however, is not technological determinism. Vattimo's (1993, 214) notes on technology and postmodernity are instructive: "what concerns us in the postmodern age is a transformation of (the notion of) Being as such—and technology, properly conceived, is the key to that transformation."

Medium theorists have long argued that technological media privilege particular social epistemes and identities while weakening others. Harold Innis's (1950) famous formulation, *Empire and Communication*, pointed out that written media extend administrative control through time, while oral traditions extend it temporally. Media thus propose conditions of organization that are realized through societal interactions. Marshall McLuhan's medium theory focuses on how media shape individual and societal experiences. At an individual level, "hot" media like radio and print are authoritative and do not allow for much audience participation, but "cool" media like television and telephone do allow for interaction and participation. McLuhan would probably argue that information networks are cool interactive media, albeit where the possibilities of conflict and cooperation are endless as we come together into a global village (McLuhan and Powers 1989). This may be explained as follows: "The alphabet (and its extension into typography) made possible the spread of power that is knowledge and shattered the bonds of tribal man, thus exploding him into an agglomeration of individuals. Electric writing and speed pour upon him instantaneously and continuously the concerns of all other men. He becomes tribal once more. The human family becomes one tribe again" (McLuhan 1964).

Benedict Anderson, while not a medium theorist, is appreciative of the transformative features of media. The spread of printed vernacular languages, as opposed to Latin, when printing began helped to form notions of nationalism and the "imagined community" of a nation-state:

> These print-languages laid the basis for a national consciousness in three distinct ways. First and foremost, they created unified fields of exchange and communication below Latin and above the spoken vernaculars. . . . Second, print-capitalism gave a new fixity to language, which in the long run helped to build that image of antiquity so central to the subjective idea of the nation. . . . Third, print-capitalism created languages-of-power of a kind different from the older administrative vernaculars" (Anderson 1983, 44–45).

Technology does not determine politics but with capitalism and, what Anderson calls, fatality or preexisting conditions, technology shapes the rise of nation-states and nationalism. Technology helps modernizing Europe organize territory and time.

Deibert (chapter 5) extends medium theory and Anderson's analysis to argue that the kind of collective images that information networks or hypermedia privilege differ from authoritative nation-state oriented images of the past. Ideas of security centered around nations or states are unlikely to endure in interconnected information networks. He notes the rise of "network security" in which "the primary 'threat' of the Internet is the potential for systems 'crash,' loss, theft or corruption of data, and interruption of information flows. The primary object of security is *the network*" (131).

Gilpin (1981) had argued that developments in military technology allowed states to not think of territorial expansion as the only means and end of power. However, physical territory itself, as epitomized geographically in nation-states, continued to be of importance. Deibert and others are now positing constitutive contexts where territoriality no longer governs human interaction. The world of hyperspace challenges the idea of territorial space as the only kind of space, especially defined by nation-states. Ruggie and Castells advocate looking at "space of flows" in information networks along with "spaces-of-places" that existed earlier.

The preceding analysis postulates that each epoch's interactions are in part proposed and molded by its technologies. Information technology networks in particular show how the collective social epistemes are shifting away from hierarchical authoritative contexts privileging nation-states. Interconnected networks may flatten hierarchies, or transform them altogether, into new types of spaces where territoriality itself becomes extinct.

Luke (1989) offers an alternative view. While discarding the linear perspectivism offered by modernity, he is less sanguine about empowerment of marginal actors. For him, "informational modes of production" lead to (24) "completely commodified communication" (much like Comor in chapter 7). Combining cultural theory (Horkheimer and Adorno), Semiotics (Barthes, Baudrillard) and Marxian theory, he notes (48): "The power exercised in nonlinear, screenal space, however, is more puzzling. It seems to require continuous coproduction by those with access to behind the screens and those without access before the screens. Power here is essentially seductive, motivating its subjects with images to collaborate in reproducing or completing the codes' logic or sequence at their screens. Individuals recreate themselves continuously in the permissive coding of individual self-management. The institutional leadership of informational society recognizes that 'rebelling' within such screenal spaces is not necessarily a serious threat to the social order."

Is information technology unique in speaking of meta-power? Braman (chapter 4) proposes a conservative, yet revealing, precedent. She likens information technology to biotechnology to show how both at their core contain genetic power that can be utilized to affect the behavior of systems through control over the informational bases of the materials, institutions, and ideas (94). Genetic power thus changes the very stuff of other forms of power. Braman's analysis of biotechnology, however unusual in a volume on information technologies, provides a fascinating contrast. She cautions us about thinking that the only technologies that create information bases to transform identities and agendas are information technologies. More importantly, that such technologies possess information bases adds a crucial element to our understanding of how meta-power works.

The constructivist turn in international relations scholarship, that supports the basis for what this volume terms meta-power, in its strongest version, is not merely supplementing, but also replacing traditional notions of power and authority. Nonetheless, it is hard to see how power based on capabilities, as in instrumental and structural variants, can be overlooked even in transformed contexts. This volume's chapters, therefore, often take into account several forms of power. The constitution of ideas, interests, and institutions is important but that should not limit us from noticing actors' capabilities within particular contexts. For Wendt (1992), while state interests may be reconstituted, they can also be taken as given in the short run. Similarly, this volume argues for noticing the changing scope of power in all three conceptualizations discussed above.

THE CHANGING SCOPE OF GOVERNANCE

Power is ultimately about capabilities, identities, and interests. Governance involves authority, concerted action, and the resultant institutions. Information networks themselves are governance networks. They allow for diffused forms of authority to emerge, for concerted action to take place, and for institutional creation or reinforcement. A major theme in this volume is how the locus of authority is shifting away from the state because of the rise of networks. Governance can hardly be uncomplicated or purely path dependent in a multi-actor, multi-issue world, in a state of flux. Governance takes place at both informal and formal levels and may be top-down, bottom-up or both. For Rosenau (1992, 4), governance is "a system of rule that is as dependent on intersubjective meanings as on formally sanctioned constitutions and charters."

This volume discusses governance and information technologies in two predominant ways. First, governance of specific issue-areas, from security to economic to cultural, is changing because of information networks. Information is deemed, in scholarship and popular opinion, to make governance less hierarchical and more plural and democratic. Second, international governance of information technologies, particularly telecommunications, may epitomize the new forms of governance arising in global politics. Therefore, governance both *involves* information technologies in particular issue-areas and it *is about* information technologies regarding the rules that shape information networks. As noted earlier, governance may also be affected by the type of media in use.

The rise of information networks thus impacts patterns of governance in three distinct ways: (1) states are no longer the only actors in technological matters globally, (2) we now speak more of technological plurality than of a technological order, and, (3) global advocacy networks, especially among underprivileged groups, are undermining the legitimacy of existing centers of authority.

From States to Multiple Actors

Whether the state fostered laissez-faire or dirigiste strategies in national technological deployment, they were explicitly or implicitly tied to considerations of national power. The state thus reflected the industrial age technological compact. Considerations of state power matched businesses' need for monopoly privilege (Viner 1948). For example, Zacher and Sutton (1996, 220) note that "there was a general assumption in most publics that any self-respecting nation owned and controlled its air transport, telecommunications, and postal industries." The national competitiveness debates noted previously may even be a throwback to the industrial era. Krugman (1994) explicitly likens them to mercantilist policies.

Dirigiste strategies increasing state power are well-known in cases such as the rise of Prussia under Bismarck, Japan with the Meiji restoration, and France's mercantilist *grand projets*. Similar considerations applied even where business was purportedly free. British industrial strength and its imperial designs went together; the East India Company is an obvious example. Industry in general received many special privileges from the state. As Polanyi (1944, 139) argues, even free trade was created: "Just as cotton manufactures—the leading free trade industry—were created by the help of protective tariffs, export bounties, and indirect wage subsidies, laissez-faire itself was enforced by the state."

Infrastructural industries such as shipbuilding and railways were especially encouraged by states. They helped the states strengthen administrative control over existing territories (domestic and colonial) and were often instrumental in opening new frontiers. Railroads proliferated in America, sometimes through state subsidies.[21] The building of the transcontinental railroad in the United States in 1869 and the Canadian Pacific transcontinental line in 1885 not only brought disparate frontiers together in these countries but their "lessons were not lost on the old empires in Asia, some of which similarly sought to use railroads to demonstrate sovereignty over remote territories and encourage economic and administrative development" (Pacey 1990, 150).

Industry in the United States was afforded enormous protection in the nineteenth century. An influential early exponent of the "infant industry" mercantilist tradition was Alexander Hamilton. Industry remained protectionist until its increasing international competitiveness finally allowed trade barriers to be lifted beginning with the late nineteenth century. "The 'American system' of moderately high tariff protection was explicitly enacted to stimulate and encourage the industrialization of the country" (Lake 1983). Industrial strength also came from state support given to scientific and engineering research beginning with the Merrill Land Grant Colleges Act of 1862. Universities specializing in applied research existed in the United States by the end of the nineteenth century (Nelson and Wright 1992, 1942). State support for this research was followed by business support through in-house research and development (R&D).

The state's role with respect to information technologies has now changed. First, states no longer solely promote technologies nationally and internationally. International organizations, advocacy groups, and powerful individuals are often involved. Examples include: technical standards promoted by organizations such as the United Nations or the European Union; competing global standards fostered by international businesses; promotion of information networks by domestic and international NGOs; and proliferating use of the Internet by individuals beyond the control of political authorities.

Second, whereas industrial age businesses looked for state protection, postindustrial businesses increasingly petition states for free trade. The difference is related to technology costs. As noted earlier, post-industrial technologies are more demanding in terms of geographical space and populations. Businesses can also increasingly ignore national regulations by offering products over the World Wide Web through electronic commerce. As the latter expands, the state will be further marginalized in international transactions. Rosenau (1990, 17) writes that technology allows "more people to do more things in less time and with wider repercussions than could have been imagined in earlier eras. It is technology, in short, that has fostered an interdependence of local, national, and international communities that is far greater than previously experienced."

This points to the diminishing importance of the state in human affairs. The issue here is not whether the state is a dominant political actor, which it is, but the extent to which its authority is undermined by competing domestic and international influences. Ruggie's (1993, 144) analysis of modern and postmodern space—roughly equivalent to the state's role in industrial and postindustrial times—is instructive: "the modern system of states may be yielding in some instances to postmodern forms of configuring space." "The distinctive signature of the modern—homonomous—variant of structuring territorial space is the familiar world of territorially disjoint, mutually exclusive, functionally similar, sovereign states" (151). Building on Jameson's notion of *postmodern hyperspace*, Ruggie writes of *multiperspectival institutional forms* that coexist with the state. Thus, for example, "the global system of transnationalized microeconomic links. . . . have created a nonterritorial 'region' in the world economy—a decentered yet integrated space-of-flows, operating in real time, which exists alongside the spaces-of-places that we call national economies" (172). For Rosenau (1990, 181–209), "the evolution of a multi-centric world has deteriorated the automatic authority in the past granted to state instruments. While the state may still possess instruments of coercion, its legitimacy and authority may be declining."

Many authors in this volume reveal a concern for the changed nature and role of the state, both with respect to specific issue-areas and regarding telecommunications. In terms of issue-areas, Deibert (chapter 5) shows that state attempts to regulate security from its viewpoint are no longer sustainable. Information technology does not serve either the purposes of the state or that of the nation. Another collective image of security gets favored where the "network itself is the object or referent of security" (129). This includes data security and information of importance to firms and consumers, important actors in the current global political economy. Kim and Hart (chapter 6) show that the state now plays second fiddle to global business. Comor (chapter 7) not only deals with the importance of what may be termed "private authority" in global capitalism, but also with the commodification of everyday lives that furthers global consumption patterns.[22] Comor differs from others in this volume, in that state-centric or capitalist hegemony are seen to be a false dichotomy. Instead, Comor speaks to the major theme dealing with a focus on both the changing and ongoing characteristics of power in the emerging global political economy.[23]

The changed role of the state is noted by authors with respect to telecommunication, also. Zacher (chapter 8) shows how the rules governing telecommunications were dictated by state actors alone in the nineteenth century, whereas they involve many other actors now. My chapter follows Zacher's analysis and applies it to North-South negotiations—with a twist. In a multiple actor world, the power of traditionally powerful states decreases, allowing weak states to effect a few favorable outcomes. Finally, McDowell (chapter 11), uses a term coined by Barry Buzan, to note the presence of "unlike units" in global telecommunications governance which include states, sub-state actors and international organizations.

FROM ORDER TO PLURALITY

During the industrial era, technology facilitated a sociopolitical order. The notion of this order, a set of streamlined circumstances facilitated by the extant technologies, is implicit in most writings. Technocentric ideas of progress informed by the Enlightenment revolve around this notion (Meltzer et al. 1993, 1995). At the other end of the spectrum, Marxian political economy views social relationships as being ultimately determined by technology. Either way, it is technology's exogenous role that leads to the creation of an order.

Classical and neoclassical political economy implicitly refer to an order. Adam Smith's views on the invisible hand and division of labor are examples, both dependent on available technology. Take the example of Smith's pin factory. "As every individual, therefore, endeavors as much as he can to employ his capital in support of domestick industry, and so as to direct that its produce may be of greatest value; every individual necessarily labours to render the annual revenue of the society as great as he can" (1976/1776, 456). This elegant world view derives its sanction from the prevalent moral philosophy, that promoted the virtue of individual work applied to greater good. The poem "Fable of the Bees," where each bee contributes to the hive, influenced Smith's views on division of labor and the resultant order. Such views culminate in neoclassical economics with its conception of general equilibrium and perfect competition. Landes (1980, 115) represents the peculiar economics of technology resulting from this world view: "Invention may follow genius, but production follows demand."

The sociocultural context of the modern Enlightenment beliefs also help us understand 'technological order.' Not only could rational human beings emancipate themselves from the ills of the past, but also ensure a certainty and societal progress hitherto unknown. It leads Habermas to speak of a "project" of modernity which resulted in concerted efforts "to develop objective science, universal morality and law, and autonomous art, according to their inner logic" (quoted in Ruggie 1993, 145). For Ruggie, an important manifestation of the order was the "single-point perspective" that elevated the idea of viewing art from a singular and fixed viewpoint. "What was true in the visual arts was equally true in politics: political space came to be defined *as it appeared from a single fixed viewpoint.* The concept of sovereignty, then, was merely the doctrinal counterpart of the application of single-point perspectival forms to the spatial organization of politics." It is in this context that Deibert (chapter 5) argues that security can no longer be examined purely from the singular vantage point of the state.

It is also important to understand how this political space avoided conflict, anarchy or disorder. The state being the dominant political actor, it either actively promoted a particular technological viewpoint (as in planned economies) or became the venue of conflict arbitration itself. Landes's (1980, 145) reference to "the kind of environment that generates novelty" is a synonym for the sociopolitical relations defined by the state. Whether we come from a liberal or a radical perspective, the state's role in carving a technological order cannot be denied.

The scenario changes in the post-industrial era of information networks precisely because of the fragmentation of socioeconomic life at micro and macro levels. The elegance of an order driven by the state is replaced by actor multiplicity who at times demand state intervention, sometimes run parallel to state goals, sometimes have nothing to do with the state, and at other times directly clash with the state. Deibert (1997, 205) eloquently conveys this complexity by citing postmodernism:

> Postmodern notions of "decentered" selves, pastiche-like, intertextual spatial biases, multiple realities and worlds, and fragmented imagined communities "fit" the hypermedia environment where personal information is dispersed along computer networks and privacy is rapidly dissolving, where disparate media meld together into a digital intertextual whole, where digital worlds and alternative realities are pervasive, and where narrowcasting and two-way communications are undermining mass "national" audiences and encouraging nonterritorial "niche" communities.

This context makes Rosenau (1990, 193) refer to the "advent of the multi-centric world and the concern for its actors for realizing autonomy." This concept of actor autonomy can be related to their will for empowerment. No longer do these actors wish to be part of an order defining their existence. At the grassroots level, the formation of social movements is related to their desire to be free of this order. Brecher and Costello (1994, 7–8) note how the "struggles against the New World Economy have brought about seemingly improbable alliances of environmentalists and labor unions; farmers and public health activists; advocates for human rights, women's rights, and Third World development; and others whose interests were once widely assumed to conflict." These struggles, whose concerns may be local, are often global in scope. The women's conference in Beijing in 1995 attracted about 40,000 attendees of which nearly 39,000 were from NGOs (*The Wall Street Journal* August 24, 1995, A17). Groups protesting WTO trade talks starting in Seattle on Novermber 30, 1999 and the World Bank meetings in Washington, D.C. on April 16, 2000, included environmentalists, consumer groups, human rights groups, farmers and peasants including the Zapatistas, trade unions, minorities, and religious groups. The positions taken in this volume with respect to NGOs (Litfin, chapter 3), individuals (Rosenau, chapter 11), and societal actors (Aronson, chapter 2) are consistent with the actor empowerment argument.

What happens to global governance processes when the goals of multiple actors in global politics conflict with those of others, including those of traditionally powerful states? While the process of this conflict resolution or escalation is just emerging, it is not always settled according the dictates of a state fiat alone as used to be the case. When coercion does not resolve conflict, bargaining plays a major role, as we are beginning to witness in the case of most clashes involving individuals, social movements, states and other transnational actors. Aronson (chapter 2), Zacher (chapter 8), and Singh (chapter 10) provide an outline of some of the bargaining processes underlying global politics.[24]

However, the presence of bargaining does not mean that the efforts of these non-state actors are always successful. Weak actors are often pitted against other actors who occupy dominant socioeconomic status, such as the state or transnational businesses. But, it is important that we view the conflict not just in terms of winners and losers but also in terms of the process itself which continues to instruct us on the emerging forms of authority relations. It is akin to Ruggie's (1993, 155) analogy of trade fairs in medieval Europe. "In no sense could the medieval trade fairs have become substitutes for the institutions of feudal rule. Yet the fairs contributed significantly to the demise of feudal authority relations." Nonetheless, Comor (chapter 7) and McDowell (chapter 9) do warn here that even as state hegemony declines, another hegemony, that of global capitalist businesses, is taking its place. Similarly, Luke (1989, 51) calls attention to "the flow of elite control, mass acceptance, and individual consent in a new informational social formation—the 'society of the spectacle.'"

To summarize, dominant technological relations resulting in an order often defined by the state in the industrial era are now being replaced by the multiplicity of forces involved in technology adoption or their resistance. Often the goals of these multiple actors are tied to their desire for autonomy. But information technology also reveals the fragmentation of individual and group lives. Taken together, technological pluralism may be replacing the erstwhile technological order.

FROM AUTHORITY TO ADVOCACY

The strength of the industrial era state-blessed technological order lay in its legitimacy, the latter intricately tied with the legitimacy of state instruments. In the Weberian (1968, 946) sense, legitimacy or domination is a "situation in which the manifested will (*command*) of the *ruler* or rulers is meant to influence the conduct of one or more others (*the ruled*) and actually does influence it in such a way that their conduct to a socially relevant degree occurs as if the ruled had made the content of the command the maxim of their conduct for their sake." Legitimacy in the modern (industrial) era is maintained through "a system of rational rules" which "finds its typical expression in *bureaucracy*" (Weber, 1968, 954).

The basic point is this: a technological order existed because powerful interests were legitimized through state instruments. The politics of technology in the industrial era are often the politics of the construction of this legitimacy.[25] The process of obtaining privileges, such as property rights or trade protections, from the state can thus be viewed as a construction of this legitimacy (North 1981; Tilly 1985).

The politics of legitimacy construction work differently in a multi-centric world of information networks. States are supplemented by other actors in the process of this construction and the notion of legitimacy itself is weakened when actors at various levels joust for control and influence. In a world of technological plurality, with networks empowering various actors, it is more appropriate to conceptualize technological advocacy than legitimacy. Legitimacy, even when it rests on a narrow support base, implies domination and obedience from the populations. In technological pluralism, competing or multiple technologies often have distinct, competing, or intersecting bases of support. The competing technological agendas—whether put forth by the Wintelist strategy (Kim and Hart, chapter 6) or NGOs (Litfin, chapter 3)—can be better viewed in terms of technological advocacy. Where such advocacy strengthens in constituent support, it may be described as authoritative advocacy. Whether this is grounds for describing it as legitimate is debatable. Nonetheless, texts speaking to advocacy politics with respect to information technologies keep increasing.[26]

Overall, in moving from an industrial to a post-industrial society, we are witnessing a shift in governance from state to multiple actors, and from a technological order to plurality and increased autonomy of actors involved. Thus, instead of technology helping to determine state legitimacy, increasing advocacy by different groups may be distinguished as nonauthoritative or authoritative depending on the bases of support.

CONCLUSION

Technology not only helps to shape its own circumstances, but our own understanding of technology is also tied up in them. Thus, there was a connection between industrial technology and the nation-state and also between industrialization and the Enlightenment and technocratic beliefs in progress.[27] The latter led to instrumental notions of power. But, as industrialization also created masses of urban poor along with wealthy capitalists, structural ideas of power began to supplement, and at times, contradict instrumental understandings. Utopian socialists, Marxists, and people like Thorstein Veblen contributed to this intellectual project.

While the old ones are still extant, information technology is helping to bring about new politics and new intellectual configurations. These include the following. First, the nation-state must now confront, support, or coexist with other international actors. Second, our understanding of instrumental and structural powers, both resting on notions of capability, must be reconfigured to account for digital technologies. Power may now be accruing to NGOs, international organizations, businesses, transnational social movements, and to weak nation-states. These further challenge our understanding of the nation-state and its modus operandi. Third, most importantly, information technologies are helping to reconstitute identities and issues. If preferences of actors are defined by how they interact, information networks are fundamentally interactive. Similarly, identity formation is undergoing a shift. These actors and their interactions are also reconstituting time and space. The temporal shift comes from the speed of human interactions coming from growing network interdependencies, impacting everything from military readiness to global electronic economic transactions and cultural flows. Spatially, cyberspace must coexist with territorial space. Security, economics, culture are transformed as a result.

Finally, the promises and perils of information technology need to be understood with reference to digitization and cost dynamics. Consider the following statement by Mathews (1997, 54): "Technology is fundamental to NGOs' new clout. The nonprofit Association for Progressive Communications provides 50,000 NGOs in 133 countries access to the tens of millions of Internet users for the price of a local call. The dramatically lower costs of international communication have altered NGOs' goals and changed international outcomes." What Mathews notes in terms of NGOs and information technology applies equally well to all international actors and their issues.

This volume's authors attempt to grapple with the effect of information technology summarized above. While information technologies might be responsible for fundamental transformations, the latter so far are not clearly understood in international relations scholarship. Rosenau (1997, 17) writes that technology "has profoundly altered the scale on which human affairs take place" but he goes on to add that "students of global politics have not begun to take account of transformations at work within societies." Daniel Bell (1980, X) calls it "an extraordinary transformation, perhaps even greater in its impact than the industrial revolution of the previous century" and goes on to note "that to the extent that we are sensitive, we can try and estimate the consequences and decide which policies we should choose, consonant with the values we have, in order to shape, accept, or even reject the alternative futures that are available to us."

To be sure, the effects of the so-called information revolution are heavily debated,[28] but there is no consensus. Particular effects in specific issue-areas and subfields are least understood. Many scholars also caution us against reading too much into such effects. Keohane and Nye (1998, 82) write that while information technologies will tear down old hierarchies and shape new identities, "[P]rophets of a new cyberworld, like modernists before them, often overlook how much the new world overlaps and rests on the traditional world in which power depends on geographically based institutions."

Walt Whitman's questions, still unanswered, are as relevant now, as when he penned them. We are still trying to figure out how and if nations are communing, if a global heart is developing, if humanity is forming en-masse, and which tyrants are trembling and crowns are dimming. One hundred thirty-five years after Whitman's poem, and on the threshold of a new millennium, words by Keohane and Nye in the last paragraph are eerily similar to those of Whitman's when he notes that "[N]o one knows what will happen next." This volume, hopefully, provides a few reasonable conjectures.

NOTES

I would like to thank Mary Beth Melchior, Chuck Johnson, and the authors of this volume (especially Sandra Braman, Edward Comor and Jim Rosenau) for comments on earlier drafts of this paper.

1. For histories of the International Telegraph Union, later International Telecommunication Union or ITU, see Codding (1972) and Headrick (1991).

2. In the works just mentioned, no author deals specifically with transformational issues. Most of the works deal with the creation and sustenance of the global telecommunications regime, which specifies the "principles, norms, rules and decision-making procedures" (Krasner 1985, 4) for international actors by either attributing it to states or to businesses and international organizations. Krasner takes a statist line while Aronson and Cowhey, Sandholtz, and Cowhey take a neo liberal position. Zacher with Sutton synthesize neo-liberal and neo-realist analyses. Gilpin is not explaining regimes, but his analysis remains limited to state-power and the effects of technology on this.

3. This section borrows from Sheth and Singh (1994)

4. Behind such optimism are the technological processes underscored in this subsection. "The harmonizing effect of being digital is already apparent as previously partitioned disciplines and enterprises find themselves collaborating, not competing" (Negroponte 1995, 230).

5. Costs were as high as $200,000 per MIPS (millions of instructions per second) on a mainframe when introduced. They were less than $100 per MIPS on a PC by 1995 and were expected to decline to a few dollars per MIPS by 2010. Figures quoted from Sheth and Singh (1994, 4–5). Apple computer has now released the first home computer that processes above the supercomputer threshold of 1 billion instructions per second (http://www.apple.com accessed November 10, 1999).

6. For an excellent introduction to the cost economics and the marketing issues facing information products, see Shapiro and Varian (1999).

7. Milner (1997), Keohane (1984), Gilpin, (1981). Gilpin's early works (1962, 1968, 1975) are the first to write of this explicitly. Gilpin (1981) is a later example where the cost-benefit calculations of power holders, including imperial reach, are related to the technological capabilities of states.

8. Machlup (1962) noted that the rate of growth of the information sector in the economy was much faster than that of agriculture or industry. The implication was obvious: to produce high growth rates, the country needed to boost the information sector and

its employment. Porat's (1977) work later developed categories of information occupations for national income accounts.

9. Histories of Pentagon's support for AT&T may be found in Schiller (1982) and Horwitz (1989).

10. There were dissenters. Many objected to the supply-centric arguments. Aronson (1992) characterized this as the "field of dreams" approach that posited a "build it and they will come" view of the infrastructure. Krugman (1994) was a vociferous critic of what he called the "competitiveness obsession." Nations are not corporations, he argued, and should stop trying to compete with each other. While correct on the economics, Krugman overlooked politics. Nations compete not because they see themselves as corporations but because of concerns about national power, well-known from Thucydides to Kennedy (1987). This is not to say that competitiveness obsessions are desirable. And, as Deibert points out in chapter 5, notions of 'national power' may themselves be history soon.

11. This is not to dismiss suasion altogether. Keohane and Nye (1998) note that attraction of actors toward mutually desired outcomes, what they term 'soft power,' is important for the world in which interdependence is deepening due to information technologies.

12. Instrumental notions of technology in fact have their origins in Western liberal thought (Meltzer et al. 1993, 1995). The idea that technology is intimately tied to empowerment of the less privileged epitomizes such thinking. An extreme version is the notion that all social problems can be reduced to technological ones. For a critique, see Sarewitz (1996).

13. For an application of Coxian framework to information technologies, see Sinclair (1999). He argues that the financial credit rating structure, sustained by a global electronic network, regulates the behavior of organizations and individuals.

14. What liberal political thought is to instrumental notions of technology, radical political thought may be to its structural variant. Most structural variations of technology, even the neoclassical ones, borrow from radical scholarship. (See Schumpeter, 1939; Archibugi and Michie, 1997). Alternative structural notions embedded in realist analysis, starting with Thucydides's maxims about the weak suffering what they must, have not espoused much literature examining technological questions. Gilpin's work, mentioned earlier, is an exception.

15. According to Polanyi (1944, 132), laissez-faire was more of an organizing principle than a free for all system.

16. Also see North (1990, 1981).

17. Well-known works include Finnemore (1998), Keck and Sikkink (1998), Katzenstein (1996), Biersteker and Weber (1996), Ruggie (1993), Wendt (1992), Onuf (1989). Postmodernists and gender theorists, whose work overlaps with this tradition, include Weber (1999), Peterson (1997), Walker (1993), Enloe (1993), Der Derian and Shapiro (1989).

18. For an application to communication media, see Singh (1999).

19. Comor (chapter 7) borrows from Berger and Luckmann to point out how the capitalist institution of consumption gets socially constructed, or how individuals are socialized to consume.

20. A recent *The Wall Street Journal* (September 12, 2000: B1) in a news story titled "Now You, Too, Can Be Spy," noted that spy images available only to the military earlier are now available to everyone at low prices.

21. Fishlow (1965) provides an excellent introduction to how railways helped to meet the demands of the antebellum U.S. economy.

22. The emergence of private authority in global politics is now increasingly noted. See Cutler et al. (1999). Cutler's (1999) view comes close to that of Comor's here and Comor (1994). Whereas at one time commercial interests required the state to further their interests thus leading to 'public law,' increasingly the commercial interests want international private law which places their activities outside the purview of state instruments.

23. Intellectual antecedents to Comor's approach may also be located in Adorno (1991), Horkeimer and Adorno (1972), Gramsci (1971), and Luke (1989).

24. Separately, I have argued (Singh 2000b) that bargaining increasingly favors weaker actors in a multiple-issue, multiple-actor world as their alternatives become better.

25. This discussion draws from Borgatta and Borgatta (1992, 1095–1099) on legitimacy.

26. For recent examples, see Singh (2001), Keck and Sikkink (1998), Mathews (1997), Deibert (1997, 157–164), Castells (1997), Rosenau (1990), Smith and Guarnizo (1998), Brecher and Costello (1998), Luke (1989).

27. See Simpson (1995) for the connection between technology and modernity and Meltzer et al. (1993) for technology and liberal democracy. Kass (1997, 22), while not agreeing that instrumental beliefs are unproblematic, notes in the American context that "the preservation of our liberties, no less than our general welfare, has been tied on more than one occasion to American engineering, rational planning, and methodological social organization; I refer in particular to the Second World War." See Ezrahi et al. (1994) for how progress-oriented beliefs in technology are giving way to pessimism in postmodern times.

28. Well-known works, apart from ones listed earlier, include Castells (1996, 1997, 1998), Sapolsky et al. (1992), Pool (1990), Beniger (1986), Pavlic and Hamelink (1985), Rogers (1983), Nora and Minc (1980), Porat (1977), Bell (1973).

REFERENCES

Adorno, Theodor W. 1991. *The Culture Industry: Selected Essays on Mass Culture*. London: Routledge.

Anderson, Benedict. 1983. *Imagined Communities*. London: Verso.

Archibugi, Daniele, and Jonathan Michie, eds. 1997. *Technology, Globalization and Economic Performance*. Cambridge: Cambridge University Press.

Arquila, John, and David Ronfeldt. 1997. *In Athena's Camp: Preparing for Conflict in the Information Age*. Rand: National Defense Research Institute.

Aronson, Jonathan D. 1992. Telecommunications Infrastructure and U.S. International Competitiveness. In *A National Information Network: Changing Our Lives in the 21st Century.* Institute for Information Studies. The Aspen Institute.

Aronson, Jonathan D., and Peter F. Cowhey. 1988. *When Countries Talk: International Trade in Telecommunications Services.* The American Enterprise Institute.

Bell, Daniel. 1973. *The Coming of Post-Industrial Society: A Venture in Social Forecasting.* New York: Basic Books.

———. 1980. Introduction. In Simon Nora and Alain Minc. *The Computerization of Society: A Report to the President of France.* Cambridge: MIT Press.

Beniger, James R. 1986. *The Control Revolution: Technological and Economic Origins of Information Society.* Cambridge: Harvard University Press.

Berger, Peter L., and Thomas Luckmann. 1966. *The Social Construction of Reality: A Treatise in the Sociology of Knowledge.* New York: Anchor Books.

Best, Michael H. 1990. *The New Competition: Institutions of Industrial Restructuring.* Cambridge: Harvard University Press.

Biersteker, Thomas J., and Cynthia Weber. 1996. *State Sovereignty as Social Construct.* Cambridge: Cambridge University Press.

Borgatta, Edgar F., and Marie L. Borgatta, eds. 1992. *Encyclopedia of Sociology*, Vol. 3. New York: Macmillan.

Braman, Sandra. 1995. Horizons of the State: Information Policy and Power. *Journal of Communication.* 45: 4–24.

Brecher, Jeremey, and Tim Costello. 1998. *Global Village and Global Pillage: Economic Reconstruction from the Bottom Up.* Boston: South End Press.

Bruce, Robert R., Jeffrey P. Cunard, Mark D. Director. 1988. *The Telecom Mosaic: Assembling The New International Structure.* Fromme, Somerset: Butterworth Scientific for International Institute of Communications.

Castells, Manuel. 1996. *The Information Age: Economy Society and Culture. Volume I. The Rise of the Network Society.* Oxford: Blackwell.

———. 1997. *The Information Age: Economy Society and Culture. Volume II. The Power of Identity.* Oxford: Blackwell.

———. 1998. *The Information Age: Economy Society and Culture. Volume III. End of Millenium.* Oxford: Blackwell.

Codding, George Arthur. 1972. *The International Telecommunication Union: An Experiment in International Cooperation.* Leiden: E.J. Brill.

Comor, Edward, ed. 1994. *The Global Political Economy of Communication: Hegemony, Telecommunication and the Information Economy.* London: Macmillan Press Ltd.

Cowhey, Peter F., and Jonathan D. Aronson. 1993. *Managing the World Economy: The Consequences of Strategic Corporate Alliances.* New York: Council on Foreign Relations Press.

Cowhey, Peter F. 1990. The International telecommunications Regime: The Political Roots of Regimes for High-Technology. *International Organization.* 34: 169–199.

Cox, Robert W. 1987. *Production, Power, and World Order: Social Forces in the Making of History.* New York: Columbia University Press.

Cutler, Clare A. 1999. Locating 'Authority' in the Global Political Economy. *International Studies Quarterly.* 43: 59–81.

Cutler, Clare A., Virginia Haufler and Tony Porter, eds. 1999. *Private Authority and International Affairs.* Albany: State University of New York Press.

Deibert, Ronald. 1997. *Parchment, Printing, and Hypermedia: Communication and World Order Transformation.* New York: Columbia University Press.

Der Derian, James. 1990. The (S)pace of International Relations: Simulation, Surveillance, and Speed. *International Studies Quarterly.* 44: 295–310.

Der Derian, James, and Michael J. Shapiro, eds. 1989. *International/Intertextual Relations.* Lexington: Lexington Books.

Duch, Raymond M. 1991. *Privatizing the Economy: Telecommunications Policy in Comparative Perspective.* Ann Arbor: University of Michigan Press.

Enloe, Cynthia. 1993. *The Morning After: Sexual Politics at the End of the Cold War.* Berkeley: University of California Press.

Executive Office of the President. 1997. *Cybernation: The American Infrastructure in the Information Age: A Technical Primer on Risks and Reliability.* Office of Science and Technology Policy.

Ezrahi, Yaron, Everett Mendelsohn, and Howard P. Segal. 1994. *Technology, Pessimism, and Postmodernism.* Amherst: University of Massachusetts Press.

Fine, Ben, and Laurence Harris. 1979. *Rereading Capital.* New York: Columbia University Press.

Finnemore, Martha. 1999. Norms, Culture and World Politics: Insights from Sociology's Institutionalism. *International Organization.* 325–348.

Fishlow, Albert. 1965. *American Railroads and the Transformation of the Ante-Bellum Economy.* Cambridge: Harvard University Press.

Foucault, Michel. 1970. *The Order of Things: An Archeology of the Human Sciences.* New York: Vintage.

——. 1977. *Discipline and Punish: The Birth of the Prison.* London: Allen Lane.

Fuchs, Gerhard. 1993. *ISDN: 'The Telecommunications Highway for Europe after 1992?' or 'Paving a Dead End Street?' The Politics of Pan-European Telecommunications Network Development.* Max-Planck-Institut fur Gesellschaftsforsorschung Discussion Paper 93/6.

Geertz, Clifford. 1973. *The Interpretation of Cultures.* New York: Basic Books.

Gereffi, Gary. 1995. Global production systems and third world development. In Barbara Stallings, ed. *Global Change, Regional Response: The New International Context of Development.* Cambridge: Cambridge University Press.

Gill, Stephen, and James H. Mittelman, eds. 1997. *Innovation and Transformation in International Studies.* Cambridge: Cambridge University Press.

Gilpin, Robert. 1962. *American Scientists and Nuclear Weapons Policy.* Princeton: Princeton University Press.

——. 1968. *France in the Age of the Scientific State.* Princeton: Princeton University Press.

——. 1975. *U.S. Power and the Multinational Corporation.* New York: Basic Books.

——. 1981. *War & Change in World Politics.* Cambridge: Cambridge University Press.

Gramsci, Antonio. 1971. *Selections for the Prison Notebooks.* London: Lawrence and Wishart.

Haas, Ernst. 1989. *When Knowledge is Power.* Berkeley: University of California Press.

Halbwachs, Maurice. 1992/1941. *On Collective Memory.* Chicago: University of Chicago Press.

Hart, Jeffrey A. 1992. *Rival Capitalists: International Competitiveness in the United States, Japan, and Western Europe.* Ithaca: Cornell University Press.

Headrick, Daniel. 1991. *The Invisible Weapon: Telecommunications and International Politics, 1851–1945.* New York: Oxford University Press.

Hobsbaum, Eric J. 1968. *Industry and Empire: From 1750 to the Present Day.* Harmondsworth, England: Penguin Books.

Horkheimer, Max, and Theodor Adorno. 1972. *Dialectic of Enlightenment.* New York: Herder and Herder.

Horwitz, Robert Britt. 1989. *The Irony of Regulatory Reform: The Deregulation of American Telecommunications.* New York: Oxford University Press.

Hudson, Heather E. et. al. 1979. *The Role of Telecommunications in Socioeconomic Development.* Report prepared for the ITU and published by Boston: Information Gatekeepers Inc.

Innis, Harold. 1950. *Empire and Communications.* Oxford: Clarendon Press.

International Telecommunications Union (ITU). 1984. *The Missing Link: Report of the Independent Commission for World Wide Telecommunications Development.* Geneva: International Telecommunications Union.

Kass, Leon R. 1993. Introduction: The Problem of Technology. In Arthur M. Melzer, Jerry Weinberger, and M. Richard Zimman, eds. *Technology in the Western Political Tradition*. Ithaca: Cornell University Press.

Katzenstein, Peter J., editor. 1996. *The Culture of National Security: Norms and Identity in World Politics*. Ithaca: Cornell University Press.

Keck, Margaret E., and Kathryn Sikkink. 1998. *Activists Beyond Borders: Advocacy Networks in International Politics*. Ithaca: Cornell University Press.

Kennedy, Paul. 1987. *Rise and Fall of Great Powers: Economic and Military Conflict from 1500 to 2000*. New York: Random House.

Keohane, Robert O., and Joseph S. Nye, Jr. 1998. Power and Interdependence in the Information Age. *Foreign Affairs*. (September/October) 77(5).

Keohane, Robert O. 1984. *After Hegemony: Cooperation and Discord in the World Political Economy*. Princeton: Princeton University Press.

Krasner, Stephen. 1991. Global telecommunications and National Power. *International Organization*. 43: 336–366.

Krasner, Stephen D. 1985. *Structural Conflict: The Third World Against Global Liberalism*. Berkeley: University of California Press.

Krugman, Paul. 1994. Competitiveness: A Dangerous Obsession. *Foreign Affairs*. (March/April): 1–17.

Lake, David. 1983. International Economic Structures and American Foreign Economic Policy, 1887–1934. *World Politics*. (July) 35(4).

Landes, David S. 1969. *The Unbound Prometheus: Technological Change and Industrial Development in Western Europe from 1750 to the Present*. New York: Cambridge University Press.

———. 1980. The Creation of Knowledge and Technique: Today's Task and Yesterday's Experience. *Daedulus*.

Larson, James F. 1995. *The Telecommunications Revolution in Korea*. Hong Kong: Oxford University Press.

Levy, Brian, and Pablo T. Spiller, eds. 1996. *Regulations, Institutions and Commitment: Comparative Studies of Telecommunications*. Cambridge: Cambridge University Press.

Lindblom, Charles E. 1977. *States and Markets: The World's Political-Economic Systems*. New York: Basic Books.

Litfin, Karen. 1994. *Ozone Discourses: Science and Politics in International Environmental Negotiations*. New York: Columbia University Press.

Luke, Timothy. 1989. *Screens of Power: Ideology, Domination and Resistance in Information Society*. Urbana: University of Illinois Press.

Machlup, Fritz. 1962. *The Production and Distribution of Knowledge in the United States.* Princeton: Princeton University Press.

Mathews, Jessica T. 1997. Power Shift. *Foreign Affairs.* (January/February) 76(1).

McDowell, Stephen D. 1997. *Globalization, Liberalization and Policy Change: A Political Economy of India's Communication Sector.* London: Macmillan Press Limited.

McLuhan, Marshall. 1964/1997. *Understanding Media: The Extensions of Man.* Cambridge: The MIT Press.

Mcluhan, Marshall, and Bruce R. Powers. 1989. *The Global Village.* New York: Oxford University Press.

Melody, William. 1997. *Telecom reform: Principles, policies and regulatory practices.* Lyngby: Technical University of Denmark.

Meltzer, Arthur M., Jerry Weinberger, and M. Richard Zinman. 1993. *Technology in the Western Political Tradition.* Ithaca: Cornell University Press.

———. 1995. *History and the Idea of Progress.* Ithaca: Cornell University Press.

Miliband, Ralph. 1969. *The State in Capitalist Society: An Analysis of the Western System of Power.* New York: Basic Books.

Milner, Helen. 1997. *Interests, Institutions, and Information: Domestic Politics and International Relations.* Princeton: Princeton University Press.

Negroponte, Nicholas. 1995. *Being Digital.* New York: Alfred A. Knopf.

Nelson, Richard R. and Gavin Wright. 1992. The Rise and Fall of American Technological Leadership. *Journal of Economic Literature.* (December) 30(4).

The New York Times. 1997. For France, Sagging Self-Image and Esprit: Liberty, Equality, Anxiety (A Special Report). (February 11): A1.

Nora, Simon, and Alain Minc. 1980. *The Computerization of Society: A Report to the President of France.* Cambridge: MIT Press.

North, Douglass C. 1981. *Structure and Change in Economic History.* New York: W.W. Norton.

———. 1990. *Institutions, Institutional Change and Economic Performance.* Cambridge: Cambridge University Press.

North, Douglass C., and Robert T. Thomas. 1973. *The Rise of the Western World: A New Economic History.* Cambridge: Cambridge University Press.

NTIA. 1985. *The NTIA Competition Benefits Report.* U.S. Department of Commerce. National Telecommunications and Information Administration.

———. 1988. *Telecom 2000 Report.* U.S. Department of Commerce. National Telecommunications and Information Administration.

———. 1991. *The NTIA Infrastructure Report: Telecommunications in the Age of Information.* U.S. Department of Commerce. National Telecommunications and Information Administration.

Nye, Joseph Jr., and William A. Owens. 1996. America's Information Edge. *Foreign Affairs.* (March/April) 75: 45–61.

O'Brien, Rita Cruise, ed. 1983. *Information, Economics and Power: The North-South Dimension.* Boulder: Westview Press.

Odell, John. 2000. *Negotiating the World Economy.* Ithaca: Cornell University Press.

Olson, Mancur. 1965. *The Logic of Collective Action: Public Goods and the Theory of Groups.* Cambridge: Harvard University Press.

———. 1982. *The Rise and Decline of Nations: Economic Growth, Stagflation, and Social Rigidities.* New Haven: Yale University Press.

Onuf, Nicaholas. 1989. *A World of Our Making: Rules and Rule in Social Theory and International Relations.* Columbia: University of South Carolina Press.

Pacey, Arnold. 1990. *Technology in World Civilization.* Cambridge: The MIT Press.

Pavlic, Breda, and Cees J. Hamelink. 1985. *The New International Economic Order: Links Between Economics and Communications.* Belgrade: United Nations Education, Science and Cultural Organization.

Peterson, Spike V. 1997. Whose Crisis? Early and Post-Modern Masculinism. In Stephen Gill and James H. Mittelman, eds. *Innovation and Transformation in International Studies.* Cambridge: Cambridge University Press.

Petrazinni, Ben. 1995. *The Political Economy of Telecommunications Reform in Developing Countries: Privatization and Liberalization in Comparative Perspective.* Westport, Conn: Praeger.

Polanyi, Karl. 1944. *The Great Transformation: The Political and Economic Origins of our Time.* Boston: Beacon Press.

Pool, Ithiel de Sola. 1990. *Technologies Without Boundaries: On Telecommunications In a Global Age.* Cambridge: Harvard University Press.

Porat, Marc Uri. 1977. *The Information Economy: Definition and Measurement.* Washington: Office of Telecommunications, U.S. Department of Commerce.

Postman, Neil. 1985. *Amusing Ourselves to Death: Public Discourse in the Age of Show Business.* New York: Penguin Books.

Putnam, Robert. 1995. Bowling Alone: America's Declining Social Capital. *Journal of Democracy.*

Rogers, Everett. 1983. *Diffusion of Innovations.* 3rd. ed. New York: The Free Press.

Rosecrance, Richard. 1986. *The Rise of the Trading State: Commerce and Conquest in the Modern World.* New York: Basic Books.

———. 1996. The Rise of the Virtual State. *Foreign Affairs.* 75: 45–61.

Rosenau, James N. 1990. *Turbulence in World Politics: A Theory of Change and Continuity.* Princeton: Princeton University Press.

———. 1992. Governance, Order, and Change in World Politics. In James N. Rosenau and Ernst-Otto Czempiel. *Governance Without Government: Order and Change in World Politics*. Cambridge: Cambridge University Press.

———. 1997. *Along the Domestic-Foreign Frontier: Exploring Governance in a Turbulent World*. Cambridge: Cambridge University Press.

Ruggie, John Gerard. 1993. Territoriality and Beyond: Problematizing Modernity in International Relations. *International Organization*. 17: 139–174.

Sabel, Charles F. 1982. *Work and Politics: the Division of Labour in Industry*. Cambridge: Cambridge University Press.

Said, Edward. 1978. *Orientalism*. New York: Vintage.

Sandholtz, Wayne. 1992. *High-Tech Europe: The Politics of International Cooperation*. Berkeley: University of California Press.

Sapolsky, Harvey M. et al. eds. 1992. *The Telecommunications Revolution: Past, Present and Future*. London: Routledge.

Sarewitz, Daniel. 1996. *Frontiers of Illusion: Science, Technology, and the Politics of Progress*. Philadelphia: Temple University Press.

Saunders, Robert J., Jeremy J. Warford, and Bjorn Wellenius. 1994, 2nd ed./1983. *Telecommunications and Economic Development*. Washington, D.C.: The World Bank.

Schiller, Dan. 1982. *Telematics and Government*. Norwood, NJ: Ablex Publishing.

Schumpeter, Joseph A. 1934. *The Theory of Economic Development: An Enquiry into Profits, Capital, Interest and the Business Cycle*. Cambridge: Harvard University Press.

Shapiro, Carl, and Hal R. Varian. 1999. *Information Rules: A Strategic Guide to the Network Economy*. Boston: Harvard Business School Press.

Schumpeter, Joseph, A. 1939. *Business Cycles; a Theoretical, Historical, and Statistical Analysis of the Capitalist Process*. New York: McGraw-Hill.

Sell, Susan. 1998. *Power and Ideas: North-South Politics of Intellectual Property and Antitrust*. Albany: State University of New York Press.

Sheth, Jagdish, and J.P. Singh. 1994. The Future of Telecommunications Services at the Local Level, In Georgia Center for Advanced Telecommunications Technology. *Telecommunications Policy in Georgia*. Georgia Institute of Technology. (October).

Simpson, Lorenzo C. 1995. *Technology Time and the Conversations of Modernity*. London: Routledge.

Sinclair, Timothy. 1999. Bond Rating Agencies and Coordination in the Global Political Economy. In A. Clare Cutler, Virginia Haufler and Tony Porter, eds., *Private Authority and International Affairs*. Albany: State University of New York Press.

Singh, J.P. 1994. "Faust or Frankenstein: Whose Life is it Anyway? An Essay on Langdon Winner's 'Autonomous Technology'." American Political Science Assn. New York. (September 4)

——. 1999. Collective Memory and Colonialism: Opera and its Gendered Politics. Paper presented at the American Political Science Association, Atlanta, Sept. 2 and at the Third Pan-European International Relations Conference, Vienna, Austria. (September 16–19, 1998)

——. 1999. *Leapfrogging Development?: The Political Economy of Telecommunications Restructuring.* Albany: State University of New York Press.

——. 2000a. "The institutional environment and effects of telecommunication privatization and market liberalization in Asia." *Telecommunications Policy.* 24.

——. 2000b. Weak Powers and Globalism: Impact of Plurality on Weak-Strong Negotiations in the International Economy. *International Negotiation.* 5.

——. 2001. Gender-based NGO Networks and Information Technologies in India." In Tim Shaw, Sandra Maclean and Fahimul Quadir. eds. *Ethnicities and Governance in Asia and Africa.* Ashgate Publishers.

——. forthcoming. *Communications and Diplomacy: Negotiating the Global Information Economy.* Basingstoke: Macmillan/St. Martin's (international political economy series).

Singh J.P., and Jagdish Sheth. 1997. "Exclusion and Territoriality in Global Telecommunications: Influence of Industrial Age State-Business Relations in the Information Age." Paper Presented at the International Studies Association Annual Convention, Toronto. March.

Sisodia, Rajendra S. 1992. Singapore Invests in the Nation-Corporation. *Harvard Business Review.* (May–June).

Smith, Adam. 1976/1776. *An Enquiry in to the Nature and Causes of the Wealth of Nations.* Indianapolis: Liberty Classics.

Smith, Michael Peter, and Luis Eduardo Guarnizo, eds. 1998. *Transnationalism from Below.* New Brunswick: Transaction Books.

Snow, Marcellus. 1988. Telecommunications Literature: A Critical Review of the Economic, Technological and Public Policy Issues. *Telecommunications Policy.* (June).

Spar, Debora. 1999. Foreign Investment and the Pursuit of Human Rights. *Challenge.* (January–February).

Spar, Debora, and Jeffrey Bussgang. 1996. Ruling the Net. *Harvard Business Review.* (May–June): 19–27.

Strange, Susan. 1988. *States and Markets.* London: Pinter.

———. 1991. An Eclectic Approach. In Craig N. Murphy and Roger Tooze. *The New International Political Economy*. Boulder: Lynne Reinner.

Tilly, Charles. 1985. War Making and State Making as Organized Crime. In Peter Evans, Dietrich Rueschemeyer, and Theda Scocpol, eds. *Bringing the State Back In*. Cambridge: Cambridge University Press.

Tyson, Laura D'Andrea. 1992. *Who's Bashing Whom? Trade Conflict in High-Technology Industries*. Washington, D.C.: Institute for International Economics.

Tyson, Laura D'Andrea, and John Zysman, eds. 1983. *American Industry in International Competition: Government Policies and Corporate Strategies*. Ithaca: Cornell University Press.

Vattimo, Gianni. 1993. Postmodernity, Technology, Ontology. In Arthur M. Melzer, Jerry Weinberger, and M. Richard Zinman, eds. *Technology in the Western Political Tradition*. Ithaca: Cornell University Press.

Viner, Jacob. 1948. Power versus Plenty as Objectives of Foreign Policy in the Seventeenth and Eighteenth Centuries. *World Politics*. 1: 1–29.

Walker, .R. B. J. 1993. *Inside/Outside: International Relations as Political Theory*. Cambridge: Cambridge University Press.

Waltz, Kenneth. 1979. *Theories of International Politics*. New York: McGraw Hill.

Weber, Cynthia. 1999. *Faking It: U.S. Hegemony in a "Post-Phallic" Era*. Minneapolis: University of Minnesota Press.

Weber, Max. 1968. *Economy and Society: An Outline of Interpretive Sociology, vol. 1–2*. Guenther Ross and Claus Wittich, eds. Berkeley: University of California Press.

Wellenius, Bjorn, and Peter A. Stern, eds. 1994. *Implementing Reforms in the Telecommunications Sector: Lessons from Experience*. Washington D.C.: The World Bank.

Wendt, Alexander. 1992. Anarchy is What States Make of it: The Social Construction of Power Politics. *International Organization*. (Spring) 46(2): 391–425.

Whitman, Walt. 1982. *Whitman: Poetry and Prose*. New York: The Library of America.

Winner, Langdon. 1977. *Autonomous Technology: Technics-out-of-Control as a Theme in Political Thought*. Cambridge: The MIT Press.

Zacher, Mark W., with Brent A. Sutton. 1996. *Governing Global Networks: International Regimes for Transportation and Communications*. Cambridge: Cambridge University Press.

CHAPTER TWO

GLOBAL NETWORKS AND THEIR IMPACT

JONATHAN ARONSON

The spread of integrated global networks is accelerating. Vast and growing quantities of information flow across these networks at ever greater speed and continually declining prices. These technologically sophisticated networks are reshaping the landscape of politics and international relations, transforming global commerce, recasting societies and cultures, and altering policy formulation and implementation. Many suggest that this is the dawn of a new information age or the onset of a world information economy. Some predict bright prospects arising from these innovations; others worry that new technologies will destroy jobs and cause a permanent "digital divide," a chasm separating rich and poor within and between countries.

The scope of change is widespread, deep, and rapid. Analysts grappling with these changes often become mired in generalities or focus on specific micro-issues, losing touch with the bigger picture. Two approaches help put things in perspective. First, historical context of the kind provided by Mark Zacher in this volume illustrate the evolution of change. Second, issues can be classified and sorted. This second approach is taken here. Three analytical distinctions help categorize issues related to changes prompted by the evolution of global networks. The goal of this taxonomic exercise is to explain in accessible, but structured, shorthand the terrain of possibilities created for policymakers, firms, and society by the new global networks while also providing a framework for theory building, not new theory.

THREE DISTINCTIONS

Three distinctions are at the core of this exercise. The first distinction divides *content and conduit issues*. Many issues arise from the management, pricing, and regulation of content. The proliferation of information flowing through wireline and wireless networks and the ease of accessing and manipulating it changes how governments and firms conduct business and how individuals live. Burgeoning information flows affect the regulation and conduct of policy and commerce. They keep people informed and allow them to make their support or outrage known and thus influence society and events. As the volume of information flows climb, changes are accelerating. By contrast, conduit issues linked to the design, financing, construction, operation, maintenance, and integration of global networks are just as important but receive less attention.

Second, both content and conduit issues can be classified according to what flows in what manner over which networks. On the content side money, E-commerce, data, and ideas all flow across networks. Most of the world's *money* pulses through global networks. Banks exchange currencies. Stocks, bonds, and commodities are bought and sold without currency ever changing hands. *E-commerce* allows the sale or auction of goods over networks, even when physical delivery is still required. In addition, bits of *information* are transmitted, viewed, analyzed, and acted upon. Telephone calls, cable and satellite television programs, news broadcasts, price quotes, and sports' odds and scores all are globally available. Inherent in some information are *ideas* with the potential to change governments, firms, societies, and their people.

On the conduit side, different concerns arise depending on whether the information is received as *voice, data,* or *images* (still or moving). These distinctions are blurring with the expansion of the *World Wide Web* and the integration of information technologies. But, presently, different issues are raised depending on how information is used. Direct communications among people (telephone calls, pagers, faxes), data transmissions (databases, marketing plans, financial records, travel reservations, electronic commerce purchases) and the broadcast of images (video-streamed events, news, and entertainment) raise distinct issues. In addition, as technologies converge, new crosscutting issues emerge. Thus, in areas like video conferencing, distance learning, and interactive entertainment the voice/data/image distinctions are eroding just as the once popular FCC distinction between basic and value-added services lost its meaning.[1]

The third distinction identifies three *arenas of policy impact:* politics and policy, commerce and finance, and society and culture. First, new technologies and global networks impact the domestic and foreign politics and policies of countries and force officials to redesign regulatory approaches. Second, globalization is transforming global commerce and finance and may impel private firms and state-controlled entities to become regional and global players to stay competitive. Finally, stimulated by the explosion of the Web and the proliferation of other inexpensive forms of communications, cultures and societies are reinventing themselves at a breathtaking pace.[2]

This section looks at content issues. The following section focuses on conduit issues and the policy questions they raise. Several themes in these sections recur throughout this volume. Communication and information technologies and the networks they enable are distributing power more widely. The implications of this shift in authority may ultimately be as great as the collapse of the Soviet Union and the end of the Cold War. States remain powerful, but competing international and nongovernmental organizations and institutions also are gaining power and influence. As J. P. Singh notes in the introduction and James Rosenau reprises in the conclusion, the decentralization of power changes how power is exercised, the proliferation of global networks raises new challenges for governance, and the ubiquity of information at low cost creates far more textured and multilayered notions of identity.

CONTENT ISSUES

People do not need to know how telephones, fax machines, computers, and televisions work to use them. The miracle of modern technology is its simplicity of use. New services attract users. Falling prices spur usage. Most people believe that the content that flows over networks and its impact on governments, firms, society, and people matters most. Here, four types of flows are considered: money, E-commerce, information, and ideas.

MONEY

Most money is electronic. Currencies—bills and coins—make up a small portion of the money supply in most industrial economies. Similarly, the percentage of banking and credit card transactions that take place over the phone or online is increasing rapidly. As smart cards, debit cards, and phone cards proliferate, the physical exchange of money should decrease. This does not mean that a cashless economy is imminent, but the trend is clear. These developments are influencing politics and policy, commerce and finance, and society and culture. This section and those that follow highlight the nature and importance of the changes in these three arenas.

Politics and Policy: Control of Money. Has the explosion of electronic money enhanced or undermined governments' monetary control? Huge sums of money move from country to country and currency to currency each day. A decade ago it was estimated that the value of foreign exchange transactions dwarfed the value of global trade by a factor of fifty (Spero 1988–89). By the end of the 1990s the value of foreign exchange trades reached about $1.2 trillion each day. Many of these transactions are intra-corporate, intraday adjustments, but still the imbalance is rising.

This maelstrom of activity raises twin challenges for monetary authorities. First, on a day-to-day basis, can central bankers still manage national money supply when money can be created beyond their control and can flood or flee a currency in an instant? Most governments no longer make any serious attempt to impose currency controls. Even the imposition of draconian penalties usually fails, falling victim to the fungibility of money. Central bankers are beginning to grapple with these issues, but so far they have downplayed their significance and asserted that they are still in control. Second, do the volume and velocity of money changing hands make it more likely that a global financial crisis could sweep from currency to currency causing a global financial meltdown? From the runs on sterling in the 1960s which began the slow death of the Bretton Wood system to the Asian financial crisis in the late 1990s, central banks, finance ministries, and the International Monetary Fund (IMF) labored to stabilize the global monetary scene. Still, crises provoked finance ministers and political leaders from London to Kuala Lumpur to blame foreign speculators for their woes.[3] Politicians like to pass the buck to avoid blame for their own mismanagement. Still, there is a growing concern that the system as managed by the IMF, if not the individual speculators, bears some responsibility for recurring crises.

Commerce and Finance: Global Disbursement and Payments. Who wins and who loses when new payment possibilities allow individuals and firms to think and act globally? Money transfers over networks make it easier to travel or stay at home. The ritual stop at the bank for cash and traveler checks before a foreign trip is no more. Visa and American Express are accepted worldwide and cash in any national currency can be withdrawn from ATMs anywhere. Similarly, direct deposit of paychecks and payments of bills are fast becoming the norm, not the exception. The location of payer and payee is irrelevant. National borders and the type of currency do not matter. The ease of money transfer and erosion of barriers also push firms, even smaller ones, to think and act regionally and globally. The variety of products and services readily available through the global electronic marketplace continues to proliferate. Governments are playing catch up, but mostly are staying out of the way and allowing firms to push economic globalization forward. Although globalization probably will proceed despite efforts to turn back the clock, much more work is needed to understand why certain firms and sectors prosper while others lag. (Friedman 1999a)

Society and Culture: Global Currency. Will integrated electronic markets with common currencies unite or divide peoples? As the millennium dawns there are two and a half currencies that matter. The dollar is solid. There even is talk, from Quebec to Argentina, that the dollar should be adopted as a single currency for the entire hemisphere. The Euro debuted on January 1, 1999, and despite birthing pains, promises to promote predictability and growth across a large region. The yen limps along as Japan marks nearly a decade of economic stagnation punctuated by unsuccessful stimulus packages. Its prospects faded when the Asian tigers stumbled, but it should endure in Asia. Most electronic purchases will be denominated in one of these currencies. Price arbitrage for similar products and services will occur. Consumers and firms will buy and sell products and services according to their quality and price and not the location of buyer or seller. Inevitably, national preferences and global tastes will collide creating cultures and societies torn between national and global preferences and further exacerbating the Jihad vs McWorld split. (Barber 1995). It is less clear, but equally important, how and to what extent global markets will impact to reinforce or erode national and ethnic identities. What does seem certain is that identity for most people and groups will be multifaceted, interlacing elements of the global and the local.

E-COMMERCE

People, money, things, information, and ideas flow across national borders. Global networks may substitute for international travel and facilitate the sale of things. Catalogue sale venders using 800 numbers are proliferating. More significantly, the emergence of the Internet and the Web set the stage for the E-commerce phenomenon. Internet users exploded from three million in 1994 to 200 million by the start of 2000 and Internet traffic was doubling every 100 days in early 2000. The speed of delivery over the Internet backbone network is increasing even more rapidly and is projected to reach 4,800 Megabits per second in 2000. Annual infrastructure investment has nearly doubled between 1996 and early 2000. Between 1995 and 1999 the average online usage per user more than doubled and could double again by 2002. E-commerce experienced a similar trajectory before slowing after March 2000. E-commerce as a percentage of U.S. GDP was essentially zero in 1995, reached 1 percent in 1999, and continues to climb (Pepper 1999). Sales of securities over the Internet exceeded $100 billion a day by early 1999 (Freedman 1999). In short, E-commerce is transforming political and policy possibilities, firms and business sectors, and the way people live and interact (Magaziner Report, 1998).

Politics and Policy: Regulatory Responsibility. How should regulators and legislators respond to the rapid expansion of electronic commerce? The Federal Communications Commission has refrained from regulating the Internet but is watching closely (Werbach 1997). The Clinton administration offered a framework for global E-Commerce in July 1997, proposing that the private sector should lead and governments should avoid undue restrictions on E-commerce. To the extent

government involvement is necessary, it should enforce a predictable, minimal-
ist, consistent, and simple legal environment for E-commerce (Clinton 1997;
Clinton and Gore 1997). Regulators believe that they should pursue greater
competition and interoperability and that there needs to be greater agreement
on how and to what extent to protect intellectual property rights. Regulators
are more hesitant to regulate content.

The Web and E-commerce also present regulators with a different set of challenges related to consumer protection because fraud is rampant and growing.
Concerns about privacy and data security, particularly for credit cards, also are
widespread. Competition for eyeballs and competition between E-commerce
providers and more established retailers is white hot. E-commerce providers con-
tend that self-regulation is the best course. Lawmakers and regulators are not so
sure. They are struggling to figure out whether and how they should intervene to
ensure that the competition is fair and robust and that everybody is connected to
everybody else. The challenges to governments are global as well as national. Indeed, any serious trade future trade negotiations will need to reach agreements
that promote rather than retard the booming global E-commerce sector.

*Commerce and Finance: Global Competition. Who will win and lose as the rapid
rise of E-commerce alters the competitive situation of individual firms and of industrial and business sectors?* Global networks erode borders, making it possible for
savvy firms to produce goods and services and compete globally. Business-to-
business E-commerce is projected to increase far faster than business-to-
consumer E-commerce. In 1999 business-to-business E-commerce reached about
$100 billion, about four times the volume of business-to-consumer E-commerce.
A quarter of American households made at least one purchase online in 1999.
By 2004 the volume of business-to-business E-commerce could dwarf the volume
of business-to-consumer E-commerce by a factor of ten to one (Pepper 1999).
The way people bank and shop is in flux. If individuals and firms do not need
to visit a bank to do business, it may not matter where the bank is located. The
death of distance that is transforming communications and commerce has
spilled over to finance and is changing the way people live.

E-commerce empowers buyers by providing them with more information
about their alternatives. Malls may suffer because it is easier to shop online.
Newspapers could suffer if they lose readers and advertisers, although advertising by many now bankrupt Web firms helped push up revenues in 1999. The
Web also could change the way goods and services are sold internationally.
American children rushed to buy the British copies of the *Harry Potter* series online before they were available in America. Future volumes of the series will be
released simultaneously on both sides of the Atlantic. German citizens may not
legally purchase Hitler's *Mein Kampf* in a German bookstore, but it is a top-sell-

ing item in Germany through AOL (Friedman 1999b). Moreover, as national boundaries become more porous, tax, trade, and intellectual property questions related to E-commerce are rising. This activity reinforces the growth of cross-national production networks that are altering the terms of competition in global markets and are transforming the structure of many industries (Kim and Hart in this volume; Borrus and Zysman 1997).

Among those who are connected, the gap between information haves and have nots is closing. Individuals and institutions have access to the same information as their brokers and merchants. Moreover, the way that more and more people are buying and selling is changing as people flock to eBay and other online auction sites. One consequence is that, unless they provide real value as infomediaries, brokers and middlemen will be squeezed out. Thus, full service stock brokers are losing market share to discount brokers like Charles Schwab and online Internet brokers like E*trade. Even the full service brokers are using the online traders to execute their trades. Customers will pay for advice but their tolerance for high transaction fees is gone. Even stock exchanges are restructuring in the face of competition from online electronic communication networks like Island which by early 1999 controlled 21.6 percent of Nasdaq shares and almost a third of Nasdaq trades (Vogelstein 1999). Similarly, computer programs like Turbotax and online tax filing are squeezing accountants; and airlines and discount online ticket providers such as Priceline.com allow customers to bypass and undercut travel agents by booking tickets online.

Society and Culture: Global Branding. Will tastes and cultures converge or remain distinct as E-commerce promotes new forms of global branding? Products like Coca Cola, Levis, and Walkman gained global acceptance long before the rise of E-commerce. In the early 1990s concern increased that as product cycles shortened only large firms with the money to mount international marketing campaigns to introduce new products would prevail across borders. E-commerce could reverse the equation by allowing customers to spread across the globe to find what they need instead of advertisers selling what they choose. (A darker view is provided by Edward Comor in this volume.) Small firms with good ideas and products, access to a large bank of computing power, and excellent international Internet connections might compete anywhere regardless of their country of origin. As trade and communications barriers fall, E-commerce provides a new, affordable way for firms to supplement their efforts to gain international recognition for their products. However, the huge expense of establishing and marketing a visible site and the advantages enjoyed by first movers, may work against innovators. In short, global branding may or may not reinforce trends already in progress and the Internet and E-commerce may play an important role in ongoing globalization.

INFORMATION

Soon almost everybody, at least in the industrial world, will be connected with everybody else in real time. The numbers are startling. In just ten years, since the first commercial Internet providers began operation in 1989, more than 200 million users are connected, including 100 million in the United States. By 1999 almost 40 percent of U.S. households were connected; 20 percent in the European Union, and 10 percent in Japan. Affordable voice, data, and video connections between people and machines are now the norm, not the exception. Any government, firm, or individual has access to more information than exists in the world's great libraries. *Encyclopedia Britannica* is accessible online for free. Interactive 900 telephone numbers, computer chat rooms, and interactive computer games now occupy so much time in America that television viewership is declining. The links are international, not just national. The Internet, Email, exploding capacity, and falling international phone rates all make it easier for governments, firms, and people to stay in touch. The implications of information abundance extend far beyond the drop in letter writing and reading.

Politics and Policy: Intelligence and Planning. Does more intelligence information translate into better policy? Governments always want to collect and analyze information that will inform their decisions. Intelligence communities want to collect as much information as possible. The information collection capabilities of modern intelligence services was demonstrated after the downing of Korean Airlines 007. Within hours, President Reagan released the conversations between the Soviet pilot who shot down the plane and his ground base. Even though the information exists, finding it in the databases and archives can be challenging. Developing efficient search routines therefore becomes imperative. However, the glut of information flows may clog the system and may not lead to better policy. There may be less room for intuition, trust, and secret understandings that were traditional instruments of the process. In short, more information may be a blessing when bureaucrats and political leaders can manage, analyze, and synthesize the data. It can be a curse when abundant information overloads or dehumanizes the decision-making process to the detriment of creativity and flexibility. Understanding how and when more information leads to better policy could become a more important area of study.

Commerce and Finance: Global Production and Marketing. Will firms be more competitive if they produce and market globally for global markets? Firms depend more than ever on information and communications to ensure their global competitive positions and long-term viability. Business strategists show that to remain dynamic firms need information to produce goods and services globally, track their operations and inventory, and market them to customers wherever they may reside. The demand for global production and marketing means that managers focus on doing business without regard to national borders specifically in the areas of trade and investment. Large firms often use information more successfully than politicians. Still, the reasons certain firms and industries adopt new communications technologies more rapidly and successfully than others and compete more effectively in global markets remains unclear. In addition, data communication networks, electronic data interchange, and improved management of information unleash new competitive possibilities for firms. Simultaneously, firms can use new technologies to meet demands at the local level. In short, a "glocal" production strategy based on ever-improving internal and external information flows can make firms more competitive. The question is, which firms and industries will benefit from such a strategy and which will falter?

Society and Culture: Instant News. How does the "CNN effect" change the way people respond to breaking events and ultimately influence the events themselves? Information is power. New technologies empower people and always threaten the establishment. This was true for the printing press which spread literacy and undermined the authority of secular and religious rules. Newspapers, telephones, television, and computers all spread information worldwide with great speed. Over time the easy access to information created a "revolution of rising expectations" but also shone light on the activities of governments and firms everywhere. Copiers and fax machines ensure that most sensitive information will leak. Leaders may not like it, but their words and actions will immediately be graded in the court of public opinion. CNN, BBC, Email, the Web, radios and telephones all spread the news worldwide in moments. Russian and Chinese leaders often learn more about what is happening during their own crises from foreign news sources and email than from their colleagues. Bill Clinton and some other leaders took the next step. They continually update policy priorities on the basis of instant polling results.

IDEAS

Ideas are not the same as information. Information provides the answers, but ideas provide the questions, dreams, and insights that reshape the world. The spread of ideas from person to person and place to place is at the core of modernization and innovation. Geography, biodiversity, and climate set the parameters. After that the speed at which ideas are transmitted and innovations are adopted is accelerated by ever-improving transportation and communication systems. The spread of new ideas is more difficult to measure than the flow of people, money, goods, or information, but their impact can be even more dramatic.

Politics and Policy: Innovation Process. Will the rapid spread of new ideas across national borders lead to meaningful policy innovation and harmonization? New technologies speed the transmission of information and the spread of ideas. But, not all new ideas are an improvement on established ones. Economists have learned a great deal about microeconomics since the 1930s, which diminishes the likelihood that the mistakes that resulted in global depression will be repeated. But, those lessons have not helped Japan break out of its malaise. Ideas for managing national economies have converged since the fall of the Soviet Union, but a chorus of criticism of the "democratic deficit" in the IMF and other international institutions persists. In the realm of communications, certain fundamental ideas favoring privatization, regulatory liberalization, and competition were rapidly adopted by many countries. Although no two regulatory authorities adopted the same approach, ideas that worked spread and found receptive officials willing to experiment with new ideas (Levy and Spiller 1996). Enough convergence and learning took place to allow the European Union to move further toward policy harmonization and to launch a common currency, something nobody predicted as recently as the 1980s.

Commerce and Finance: Self-Regulation. Will the rapid spread of new ideas change the way governments regulate and will firms effectively regulate themselves? Government recognition of the failure of micromanagement does not mean that regulators will whither away. Much will depend on how firms and individuals react to looser regulatory shackles. A balancing act is underway. Most governments now prefer markets, not regulation, to dominate, but want to ensure that privacy is maintained. Governments claim they do not want to manage content, but Congress then passed the Decency Act to try to manage access to pornography and hate sites. China stands ready to unplug broadcasters providing content unacceptable to the government. Some critics argue that ritual worship of markets is self-serving. Companies and rich individuals want fewer restrictions and taxes on their earnings. Will greed triumph, or will firms practice self-censorship and self-regulation? The answer will be mixed. Broadcasters in Asia practice self-censorship to create culturally appropriate content on a country-by-country basis. Internet portals are trying to curb spammers, but pornography and hate sites flourish. Are firms acting responsibly or are they merely wary of reregulation? Clearly, foreign firms were better corporate citizens in developing countries after the early 1970s because they learned through experience that rapacious profiteering could be hazardous to their continued operations. What kinds of carrots and sticks will induce firms to compete and self-regulate?

Society and Culture: Democracy. Does the rapid spread of information promote free-dom, democracy and market economics? In the early 1980s Ithiel Pool explored the implications of new electronic technologies for democracy and personal freedom. As Pool predicted, new technologies and the convergence of communications technologies placed new strains on freedom of speech and democracy (1983). Today, more information is available to voters on issues and candidates but the same technologies skew elections in favor of incumbents with money, name recognition, and sophisticated media strategies. The ideas that candidates espouse are only now beginning to get across. A decade ago Francis Fukuyama gained fame and ridicule when he declared that the victory of capitalism and democracy in the wake of the collapse of the Soviet Union meant the "end of history" (1989). The flow of ideas from the West suggested new possibilities and made clear what people in Communist and developing countries were missing. State suppression of ideas became more difficult. Television and movies reinforced the idea that America was the land of excitement and opportunity. Democracy and freedom are put in a spotlight, but, as Fukuyama recognized, the transition to democracy and market economics is not foreordained. Similarly, information and communications technologies in developing countries could become a means for urban elites to further distance themselves from the people or could become an important agent for societal change. It may be that the development potential of new communication and information technologies is undervalued and that, as these technologies becomes more affordable and more diverse, inequality could decline. Researchers need to go beyond slogans and consider whether and to what extent the global spread of the ideal of democracy and freedom in fact promotes democracy and freedom. Indeed, the perceived arrogance of the United States in many parts of the world and the failures to promote democracy in place of tyranny and corruption may undermine confidence in the message. Table 2.1 summarizes the preceding discussion about content issues likely to arise in today's networked economy. These issues are representative and suggestive, not all-inclusive. Still, their variety demonstrates that much work remains to be done.

CONDUIT ISSUES

Before telegraph and telephones, people carried news and letters between distant points. Fax machines, mobile phones, and computers all made it easier for people or machines to share information over great distances and among many individuals. The build-out of the infrastructure connecting these devices represents a vast investment comparable to the funds required to build the road and rail transportation systems. Although most communications specialists concentrate on the impact of information flows on various aspects of society, fewer examine

TABLE 2.1
CONTENT ISSUES

ARENAS/FLOWS	POLITICS POLICY	COMMERCE AND FINANCE	SOCIETY AND CULTURE
Money	Control of Money	Global Payments	Global Money
E-Commerce	Regulatory Responsibility	Global Competition	Global Branding
Information	Intelligence and Planning	Global Production	Instant News
Ideas	Innovation Process	Self-Regulation	Democracy

the issues related to the financing, construction, operation, and maintenance of a robust, competitive infrastructure. Until the breakup of AT&T in 1984, regulated national monopolies, some private but mostly public, provided telephone service. AT&T was not permitted to provide value-added services until after the breakup. Since then technological innovation, the fragmentation of the public networks, greater competition, and regulatory liberalization unleashed unrivaled changes in communications and information technologies (Cowhey and Aronson 1993).

VOICE

Traditionally a series of cross-subsidies existed in the pricing of telecommunications. Telecommunications services subsidized postal services. Businesses subsidized individuals. Urban callers subsidized rural users. Most strikingly, international services subsidized domestic long-distance services, which in turn subsidized local service. New competitors, technologies, and regulatory approaches combined to turn the voice communications markets on its head. Today, market competition is imperfect, but choice is growing and prices, especially for long-distance service and international long-distance calls, continue to decline toward the cost of providing services. Rate rebalancing was accelerated by the growing competitive possibilities provided by cable, satellite and Internet voice alternatives. (But, fancy technology does not guarantee successful implementation as the collapse of the Iridium satellite system demonstrates.)

Politics and Policy: Centralization of Decision-Making? How will the expansion of global communication networks change who makes the decisions and how they decide? Other chapters in this book explore how the information age is impacting government policy making. Here, it is enough to ask: Whatever happened to the plenipotentiary ambassador? Centuries ago a Renaissance ambassador upheld "his master's honour at a foreign court, aided by no more than his wit, courage, and eloquence" (Mattingly 1971, 211). Global networks allowed governments to centralize their decision-making apparatus, giving more influence to a narrow range of top-level leaders. The role of ambassadors is more social and representational than ever before. The centralization of political decision-making authority does not automatically translate into sound, efficient choices emanating from capitals. However, global networks defy easy national regulation and undermine national authority. Here the trend seems toward decentralization. Many business firms are decentralizing their decision-making process at the same time governments are moving in the opposite direction. Fortunately for good economic and commercial policy, as national and global networks proliferate, many government regulators are promoting competition instead of trying to set prices and define and defend the public interest. The problem remains, however, of how to manage or regulate global networks. Competition alone may not ensure fair, sound, and efficient service provision. For example, global satellite networks and global strategic alliances with partners in several countries may require governments to cede some authority to international institutions trying to create international rules of the road. The solution for governments may be glocal. To effectively oversee global networks may require regional or international institutions supported by national governments.

Commerce and Finance: Telemarketing. As the expansion of global communications networks allows buyers to shop anywhere, nationally or internationally, what does that change? Telephones provide opportunities for selling, buying, and soliciting contributions, as do telemarketing and Internet marketing by people and computers. Telephone and Internet marketing of goods and services is expanding. (Junk mail, faxes, and email also are proliferating.) So is home shopping. Indeed, the home shopping networks were among the first to recognize that they could show products on the air and people would call in to buy them. It now is routine to call, fax, or email in orders for everything from goods ranging from take-out food to clothes and services as diverse as airline tickets and phone sex. If prices are competitive and shopping by phone is simple, busy people will try to save time and energy by calling in their orders.

Society and Culture: Citizenship and Identity. As global communications networks evolve, how will people, particularly immigrants, define their own citizenship and identity? Until recently immigrants to the United States joined America's melting pot within a generation or so. Language, culture, and family of the old country slipped away because of the difficulty of staying in touch. Slow letters on slow boats often were sent back to uncertain destinations. People lost touch with their roots. By contrast, present day refugees and immigrants can, economically, call home or stay in touch by email and without any time lag. Communities are more likely to retain their identity and history even when they choose or are forced to leave their birthplaces. In addition, immigrants everywhere are using new technologies to reconnect to their cultures and roots. Genealogy research on the Web is popular. Heritage students are, in increasing numbers, learning the languages of their grandparents and great grandparents. There even is a move afoot for immigrants to retain the right to vote and participate in the political processes of their birth country even after they have emigrated. The move beyond the melting pot may stimulate the retention of ethnic and religious identities, but also may foster greater clashes among civilizations.

DATA/TEXT

The capacity, speed, and reach of data networks are expanding rapidly. Public and private data networks are now at the core of the operations of government and business and many individuals rely on them at work and at home. Similarly, the amount of text available online is expanding exponentially, complementing and substituting for newspapers, books, files, and libraries. The Y2K millennial scare and the ongoing specter of computer viruses are linked to the fear that they might cripple or shut down critical data networks.

Politics and Policy: Data Analysis and Speed of Response. Will global data communication networks and new information technologies allow policymakers to respond to national security situations more rapidly and efficiently? The role of information in the waging of war is changing. How different are wars in the information age? The U.S. military adapted to new realities quickly and creatively. All modern military groups are investing heavily in information and communications technologies. Their dependence on communications continues to rise. Alvin and Heidi Toffler assert that knowledge "is now the central resource of destructivity, just as it is the central source of productivity." They envision a day in the near future when "more soldiers carry computers than carry guns." They note that, in 1993, the U.S. Air Force contracted to buy up to 300,000 personal computers for its forces (Toffler and Toffler 1993, 71). The potential power of information weapons was dramatically demonstrated during the Gulf War. The military was bolstered by AWACS (Airborne Warning and Control System) which scanned the heavens for enemy aircraft and missiles and sent targeting data to allied forces from modified Boeing 707s. In parallel, J-STARS (the Joint Surveillance and Target Attack System) helped detect, disrupt, and destroy Iraqi ground forces during Desert Storm with remarkable speed and precision. But there also is danger of relying too much on new information technology. Smart networks, smart planes, and smart bombs may not substitute for soldiers on the ground.

Commerce and Finance: Telecommuting. As global data communication networks change the nature of business will the way people work change fundamentally? Global networks provide new possibilities for commerce and work. On the commercial side the whole explosion of Internet companies on the equity markets represents a bet that some of these companies will be the Microsofts, Ciscos, and Amazon.coms of the next decade. Many of these companies failed when the dot-com bubble burst in 2000. Still more will fail, but the ones that get there first could reap huge benefits. The sale of stocks, books, and airline tickets already are well advanced. The music business and real estate are in transition. The whole advertising industry is rethinking its position and newspapers and magazines must adapt or lose readers and ultimately advertising revenues. Obviously, computers and computer networks are changing the nature of work as well. It is easier and easier for people to work at home or far from the office. Mothers with young children, avid skiers, stock brokers tired of Wall Street and others often can live anywhere, even abroad, and continue on the job. Firms too can flee the urban center to the suburbs or beyond. More broadly, the networking of the world and proliferating flows of information have profound implications for the international division of labor.

Society and Culture: Education. Will global data communication networks that promote distance learning change the way we learn in the future? Can the education establishment survive the information flood? New computer technologies already have transformed the teaching and research of fields as diverse as music and classics. The new Internet technology may represent a profound challenge to education as we have known it for centuries. The explosion of new knowledge is impressive, but its distribution online may prove revolutionary. The architecture of innovation also will force changes in the structure of schools and universities. Universities will need to choose priorities and niches, and depend on interconnection to fill out their offerings. Libraries are becoming digital and global and electronic scholarly communities are emerging. Even teaching itself is beginning to change (Noam 1995, 247).[4] Distance learning is available over broadcast, cable, online, and satellite links. The University of Phoenix, highly dependent on distance and online instruction already, has the largest enrollment of any university in the United States. The transformation of educational institutions (and of publishers) will take time. Nonetheless new technologies will challenge schools and universities to reinvent themselves or risk falling victim to new, online upstarts.

IMAGE/VIDEO

If a picture is worth a thousand words, then millions of images available through the Web are valuable indeed. More spectacular still, as images move and become video a world of streaming video and countless channels of programming stretches television and movies in new directions. It is not an accident that Bill Gates is buying major photographic archives or that the value of movie libraries is soaring. Networks provide an almost limitless opportunity to deliver information to people, but the images and even more the video content needed to fill all those channels is lagging. The infrastructure required to provide those images/video on demand are vast, sophisticated, and expensive to build. The building and operation of the broadcast network raises important issues.

Politics and Policy: Credibility and Visibility. As global broadcast facilities make government actions more transparent will this encourage trust and discourage corruption? CNN, BBC, and other global broadcasters make breaking stories real. Sometimes the news providers get it wrong. Sometimes reporters are so eager to uncover wrongdoing and make their careers that they are sloppy or misleading. Still, images and video convey a credible reality that is easier for most people to grasp than spoken word (voice), print (text), or tables and graphs (data). Despite the public's revulsion to and cynicism about negative, distorted, often misleading political attack ads, they continue to be used because they work. All major politicians now have media relations experts, handlers, pollsters, and spin-meisters who try to get news skewed favorably to their politician or policies. Television is more important than retail politics in all but a few political venues. Ironically, the flood of images makes it more difficult for viewers to differentiate between reality and fiction. Historically, as Stalin demonstrated, the technology can make and alter reality and rewrite history by literally taking someone out of the picture. New computer technologies make it simple to alter, distort, or recreate the record. Conspiracy theories and "wag the dog" explanations of policy are unlikely to recede anytime soon.

Commerce and Finance: Teleconferencing and Videoconferencing. Will the spread of picturephones and videoconferencing facilities reduce the need for executive commuting and business travel? New technologies do not lead at once to expected changes. A productivity paradox arose because the introduction of computers in the work-

place did not quickly translate into productivity gains. Only after massive equipment and training expenditures take hold, does labor productivity improve
markedly. Evidence that the provision of advanced communications and information technology resulted in productivity gains was slow in coming. It is difficult to
determine when the critical threshold is reached, but the last 10 percent of the investment apparently produces most of the productivity benefits. Similarly, despite
the geometric increase in computer speed and power, the paperless office is
nowhere in sight and more trees than ever are sacrificed to the printed word. Although some firms now regularly employ teleconferencing to link key executives
who know each other already, business travel seems to be climbing not falling.
Partners in law firms or investment banks spread around the planet may save time
and effort through teleconferencing, but meetings between companies and their
clients still demand the personal touch. Over time this could change. ATMs, after
all, were rejected by most consumers when they first were introduced. People com-
plained they were too impersonal. The second time around, of course, people
started to prefer machines and their convenience to waiting in line for tellers.

*Society and Culture: Common Symbols. Will global broadcast facilities foster common
symbols that span the globe?* Images are powerful. The right picture at the right time
can catapult an individual into the limelight. Most people gain their fifteen minutes
of fame and then recede into anonymity. Each culture recognizes its own heroes.
Flags can unite a people. Even license plates may serve the purpose. In Bosnia, for
instance, one important U.S. initiative was to introduce a single license plate design
so that people in cars were not immediately identifiable as coming from one ethnic
community or another. Similarly, global broadcast networks create common global
images and symbols that spread across borders and peoples. Tragedy and triumph,
drama and soap opera, can capture the imagination or stir the emotions of people
on a level never before possible. The Gulf War, the O.J. Simpson trial, the death of
Lady Diana, or mayhem in a high school in Colorado may fascinate and appall the
entire world. Similarly the Olympics or earthquakes, famines, and other natural
disasters can generate pride or provoke generosity and empathy. The challenge is to
determine whether common symbols and images will play a positive role in transforming how people think toward each other or whether they too will get their fifteen minutes of attention and then fade away again.

THE WORLD WIDE WEB

The rapid expansion of the World Wide Web is more fundamental than the
frenzy of speculation in Internet stocks. The Web is at the locus of voice,
data/text, and image/video. All of the issues raised above come together with the

Web. Other issues arise because of the overlap and ongoing integration of information infrastructures.

Politics and Policy: Information Overload and Infosecurity. Will the rise of the Web result in information overload and the loss of privacy? When is enough information enough? Three distinct but interrelated policy issues are among those that bedevil the modern information economy. First, the superabundance of information means that some Web surfers and serious researchers spend more and more time on their computers. In addition many are watching television while online and reacting and interacting to what they are watching. Information overload is a distinct prospect and competition for "eyeballs" is intense. Second, the gold-to-junk ratio of information online is declining. When the volume of information available online is counted in terabytes, those with the most efficient search engines and search strategies are the most productive.[5] There also is a possibility that the abundance and structure of data on the Web may camouflage important information and create blind spots. Third, the protection of individual privacy is growing more difficult when everyone's spending habits, credit history, calling patterns, and communication contacts are transparent.[6] For example, technologically sophisticated "psychics" can pick up a call, see what number it comes from, identify the caller, and check their recent spending patterns in a matter of seconds. Nobody should be surprised when these "psychics" provide remarkably prescient insights into a person's life.[7] Similarly, retailers and phone operators sometimes startle callers by greeting them by name. In addition, privacy has resurfaced as a key issue in Europe and the United States, although they disagree about how it should be handled. How is it possible to balance the demands of an open transparent society necessary to guarantee freedom and still maintain a modicum of privacy? What can and should governments and other groups do to promote data security and protect privacy, or is it too late (Brin 1998)?[8] Does the partial displacement of established hierarchies by crisscrossing networks of control (as described in Ronald Diebert's chapter), mean that solutions will prove elusive?

Commerce and Finance: Intellectual Property and Standards. As the Web increases the importance of information as a strategic tool of business will instinct and intuition still be useful? Two recent books trumpet the importance of integrated information networks for business success in the future. Bill Gates focuses on the opportunities presented by online commerce and stresses that firms must use digital networks to manage their own operations (Gates 1999). By contrast, in *Information Rules*, Carl Shapiro and Hal Varian concentrate on how firms can use new digital networks as strategic tools to lock in customers and thrash competitors (1999). As Kim and Hart note later in this volume, the rise of integrated digital networks caused forward-looking firms to place renewed emphasis on protecting and managing their intellectual property. Although large questions remain about how best to protect intellectual property on the Web, in recent years the balance has tilted to favor creators of intellectual property at the expense of users of intellectual property. At both the national and international levels, large firms are increasing their efforts to protect their intellectual property so that it can be used as a competitive advantage (Lessig 1999; Samuelson 1997). During the next trade round, trade negotiators should concentrate on creating new trade rules and principles appropriate to a global information economy. They should strive to strengthen and extend the TRIPs agreement negotiated during the Uruguay Round (Barshefsky 1999).

An example of the fight to preserve dominance is the struggle by AT&T to prevent America Online and other Internet Service Providers (ISP) "open access" to AT&T cable networks to homes served by TCI and MediaOne. AT&T argues that the only way to recoup its huge investment to buy these subsidiaries is by maintaining control over the access. Oregon, backed by the ISPs, argued that this would stifle competition and innovation and raise prices to consumers (Bar et al. 1999). Similarly, new technological breakthroughs that make possible the new digital infrastructure provide ample opportunity for firms to use standards to increase their advantage over their competitors. Standards wars like those over third generation wireless standards between Qualcomm and Ericsson/Nokia and the fight over Java between Sun and Microsoft are likely to proliferate (Lemley and McGowan 1998).[9]

Society and Culture: Entertainment. How will the Web change entertainment and the way we play? Technology provides new opportunities for entertainment. Telephones, radios, televisions, video games, MP3, and computers all provide opportunities for distraction. Vast, integrated information networks are changing the nature of leisure and culture. The workings of the entertainment industry and the impact of its output on people, society, and cultures are just beginning to be appreciated. Networks of people now regularly watch, comment, obsess, and gamble about the outcome of sporting events. Chat groups, list servers, and Web sites stimulate conversations but also shape opinion about issues. Marketers regularly invade chat rooms to plant rave reviews of their products. In addition, horizontal marketing of hot entertainment products from Star Wars to Pokemon is remarkably effective and winning children's devotion. Canada and France continue to warn of U.S. cultural imperialism. France even held up the last trade round over media and cultural issues. Serious efforts by political scientists and economists to consider the economic importance and impact of the entertainment industry nationally and internationally, is however, in its early stages.

As Table 2.2 shows conduit issues are as diverse and vexing as content issues, but may not get quite as much attention.

TABLE 2.2
CONDUIT ISSUES

Arenas/Flows	Politics and Policy	Commerce and Finance	Society and Culture
Voice	Decentralization of Decision Making	Telemarketing	Identity
Data	Data Analysis & Speed of Response	Telecommuting	Education
Image	Credibility and Visibility	Teleconferencing	Common Symbols
Web	Information Overload and Infosecurity	Intellectual Property and Stardards	Entertainment

THE FORMULATION, NEGOTIATION AND IMPLEMENTATION OF POLICY

This chapter provides a wide-ranging survey of issues and questions, not a thick description of their complexities and subtleties. Still, it is worth briefly noting that the consequences of the spread of global networks unfold in stages and not all at once. Global networks are altering policy formulation, negotiations among countries, and the implementation of policy. Separating the formulation, negotiation, and implementation of policy is important because information provided by networks takes on different functions at different stages of the policy process.

    Thus, several chapters in this volume stress that the emergence of global networks and their ability to manage vast amounts of data makes it possible for governments, firms, groups, organizations, and individuals to dream of projects and consider alternative policies in new ways. Just as computers and supercomputers allowed mathematicians and scientists to attack previously impossible problems, global networks allow people to master information and use it to formulate ambitious projects. For example, networked collectors and computers now provide more data and the ability to analyze it about pollution, weather, and climate patterns than ever before. Strong data on the ozone depletion allowed countries to agree to limit certain emissions. At the same time access to global networks provides firms, nongovernmental organizations, and even individuals the possibility to compete with governments to shape or disrupt policy. For example, at the Seattle WTO Summit in late 1999, environmental and labor groups, along with dissatisfied developing countries, made it impossible for large industrial country governments to reach a compromise to launch a new round of trade negotiations. Similarly, although the vast majority of scientists now believe that global warming is occurring, a few well-funded dissenters using media networks have effectively delayed the emergence of consensus and slowed action.

Once policies and programs are formulated, global networks make it possible to implement new policies and practices on a scale and on a schedule never before possible. Firms and governments alike are spending heavily on new technologies so that they can proactively embrace change and for fear that if others get there first that they could lose out. Once the politics of national policy making are resolved and a policy is promulgated or law is passed, global networks allow for their rapid dissemination and implementation. Similarly, the introduction of new wireless networks promises cheap and rapid notification of customers as necessary. For example, if Boeing needs to update its notifications and servicing recommendations to airplane owners or Microsoft needs to update users' programs and manuals, the information can be broadcast quickly and efficiently directly to users' computers.

The undertaking of international negotiations is an example of a somewhat murky middle ground between formulation and implementation. Sometimes the policy process jumps directly from formulation to implementation. But, when governments or firms negotiate with other governments and firms, global networks may provide negotiators with an edge, a better understanding of the implications of various approaches than their counterparts. The ability to instantly access and analyze relevant information can provide a valuable advantage to negotiators. In this vein, J.P. Singh's chapter on negotiating regime change usefully points out that developing countries were more successful in the last round of WTO negotiations than generally is assumed because they agreed to adopt policies in high technology areas that they already concluded were in their interest. In return they received valuable breakthroughs from industrial countries on textiles and other traditional sectors. Similarly, U.S. negotiators were able to achieve significant success in negotiations on orbital slots because technical calculations showing the results of various approaches could be produced rapidly and shared with other negotiators. In addition, negotiators may increasingly undertake at least some sessions in ongoing negotiations, especially bilateral negotiations, via teleconferencing.

The other authors provide textured, complex, often subtle arguments that flesh out the issues raised in this overview piece. The goal here was narrower: to raise issues and questions that need to be answered. These are not the only issues; others may be equally or more important. However, the questions raised here are representative of the kinds of issues that public and private, national and international policymakers will need to address. Otherwise, the technology will drive the policy without regard to what needs to be accomplished to help people learn and prosper.

NOTES

1. The FCC stressed the basic versus enhanced distinction in its 1980 Computer Inquiry II, arguing that basic services should be regulated and enhanced services unregulated. By 1986, after the AT&T breakup, the FCC acknowledged in its Computer Inquiry III that the line between basic and enhanced services was eroding and that competition should be encouraged in the provision of all services.

2. Another possible distinction contrasts the formulation, negotiation, and implementation of policy. The information provided by networks takes on different functions at different stages of the policy process.

3. Coombs, (1976) described how central bankers organized to repel speculative attacks on sterling from 1964 to 1967. Again, in late 1997, Malaysian Prime Minister Mahathir publicly blamed George Soros and other foreign money speculators for the onset of the Asian economic crisis.

4. Christensen (1997) argues that cheap new technologies that are not immediately attractive to established customers can undermine the best run companies and advocates setting up separate subsidiaries using the new technologies to compete with the parent companies. He mentions the Internet in passing and never touches on education, but senior university administrators are worried that online education could divert revenue and force them to alter their mission and teaching methods.

5. Why would leaders commit important information to paper? Phones are easier than letters and the possibility of leaks declines if nothing is written.

6. Government security and individual privacy clash as the Clinton Administration learned during the *clipper chip* controversy. The government was caught in a no-win situation. If it can snoop on individuals it is invading their privacy. But, if it fails to gather critical intelligence and a terrorist assault succeeds, the government loses as well.

7. James Randi, the noted debunker of the paranormal, uses this example.

8. Brin (1998) concludes that somebody has to watch the watchers and hold them accountable. Karen Litfin's contribution to this volume suggests that as global satellite imagery becomes easily available to NGOs, this kind of *globalization of transparency* is taking place.

9. An Ericsson-led European consortia persuaded the EU to adopt a new standard that was not backward compatible with second-generation Qualcomm standards, and thus threatened the long-run viability of Qualcomm. After much high-level controversy, Ericsson bought Qualcomm's network business and an agreement was struck to support two standards. Qualcomm stock promptly soared.

REFERENCES

Bar, Francois, Stephen Cohen, Peter Cowhey, Brad De Long, Michael Kleeman, John Zysman. 1999. Defending the Internet Revolution in the Broadband Era: When Doing Nothing is Doing Harm. *E-conomy Working Paper 12*. Berkeley, BRIE, mimeo. (August)

Barber, Benjamin. 1995. *Jihad versus McWorld*. New York: Times Books.

Barshefsky, Ambassador Charlene. 1999. Electronic Commerce: Trade Policy in a Borderless World. Address at the Woodrow Wilson Center, Washington, D.C. (July 29).

Borrus, Michael, and John Zysman. 1997. Globalization with Borders: The Rise of Wintelism as the Future of Global Competition. *Industry and Innovation* (December) 4: 2.

Brin, David. 1998. *The Transparent Society: Will Technology Force Us to Choose Between Privacy and Freedom?* Reading, MA: Addison Wesley.

Clinton, William J. 1997. Presidential Directive on Electronic Commerce. Memorandum for the Heads of Executive Departments and Agencies. (July 1)

Clinton, William J., and Albert Gore, Jr. 1997. *Framework for Global Electronic Commerce*. (July 1)

Christensen, Clayton M. 1997. *The Innovator's Dilemma: When New Technologies Cause Great Firms to Fail*. Boston: Harvard Business School Press.

Coombs, Charles. 1976. *Arena of International Finance*. New York: John Wiley.

Cowhey, Peter F., and Jonathan D. Aronson. 1993. *Managing the World Economy: The Consequences of Corporate Alliances*. New York: Council on Foreign Relations, 164-78.

Freedman, Stuart. 1999. Remarks by Director, IBM Institute for Advanced Commerce at a Conference on The Legal and Policy Framework for Global Electronic Commerce. University of California, Berkeley. (March 5)

Friedman, Thomas L. 1999a. *The Lexus and the Olive Tree: Understanding Globalization*. New York: Farrar Strauss Giroux.

———. 1999b. Next, It's E-ducation. Op-Ed, *New York Times*. (November 17).

Fukuyama, Francis. 1989. The End of History? *The National Interest*. (Summer) 3–18.

Gates, Bill. 1999. *Business @ the Speed of Thought: Using a Digital Nervous System.* New York: Warner Books.

Lemley, Mark, and David McGowan, 1998. Could Java change everything? The competitive propriety of a proprietary standard. *The Antitrust Bulletin.* (Fall/Winter): 715–73.

Lessig, Lawrence. 1999. *Code and Other Laws of Cyberspace.* New York: Basic Books.

Levy, Brian, and Pablo T. Spiller, eds. 1996. *Regulations, Institutions, and Commitment: Comparative Studies of Telecommunications.* New York: Cambridge University Press.

Magaziner Report. 1998. First Annual Report of the U.S. Government Working Group on Electronic Commerce. Washington, D.C.: USG. (November)

Mattingly, Garrett. 1971. *Renaissance Diplomacy.* Boston: Houghton Mifflin.

Noam, Eli M. 1995. Electronics and the Dim Future of the University. *Science.* (270) (October 13).

Pepper, Robert. 1999. Global Implications of the Internet, E-Commerce and the Digital Economy, presentation by Chief, Office of Plans and Policy, FCC, to the Association of Professional Schools of International Affairs. Washington, D.C. (November 3)

Pool, Ithiel de Sola. 1983. *The Technologies of Freedom.* Cambridge: Harvard University Press.

Samuelson, Pamela. 1997. The U.S. Digital Agenda at WIPO. *Virginia Journal of International Law.* (Winter) 37:2: 369–439.

Shapiro, Carl, and Hal R. Varian. 1999. *Information Rules: A Strategic Guide to the Networked Economy.* Boston: Harvard Business School.

Spero, Joan. 1988–89. Guiding Global Finance. *Foreign Policy.* 73: 114–135.

Toffler, Alvin, and Heidi Toffler. 1993. *War and Anti-War: Survival at the Dawn of the 21st Century.* Boston: Little, Brown.

Vogelstein, Fred. 1999. A virtual stock market: Are the nation's two big stock exchanges obsolete? *US News & World Report.* 47. (April 26).

Werbach, Kevin. 1997. Digital Tornado: The Internet and Telecommunications Policy, OPP Working Paper Series 29. Washington, D.C.: FCC. (March)

World Trade Organization. 1998. Electronic Commerce and the WTO. Special Studies 2. Geneva.

PART I

THE CHANGING SCOPE OF POWER

CHAPTER THREE

PUBLIC EYES: SATELLITE IMAGERY, THE GLOBALIZATION OF TRANSPARENCY, AND NEW NETWORKS OF SURVEILLANCE

KAREN T. LITFIN

TECHNOLOGY, POWER, AND WORLD POLITICS

The impact of technological change on international politics occurs in two analytically distinct ways. From an instrumental perspective, new technologies can empower or disempower social actors—states, groups, classes, and institutions. On a more fundamental, but perhaps less visible level, technologies can influence the self-understandings and identities of social actors and perhaps even the very nature of power itself. Both sorts of arguments, instrumental and constitutive, have been made in a generalized fashion with respect to information technologies. James Rosenau's *Turbulence in World Politics*, claiming that the diffusion of information technologies has enhanced the competence of citizens and undercut the authority of states, is perhaps the most articulate example of the former sort of argument. The familiar claim that emerging forms of information-based power are supplanting material forms of power is an example of the second sort of argument (Poster 1984). In the field of international relations, the second kind of argument is most commonly grafted onto the first. For instance, the allegedly horizontal nature of information-based power has been cited as a driving force that undercut the authority of Soviet Union's centralized state apparatus (Robinson 1995). In a similar vein, Ronald Deibert (1997) argues that the diffusion of hypermedia, or new digital communications technologies, fosters a social epistemology favoring nonterritorial institutions and fragmented identities over the nation-state. Similarly, Deibert's essay in this volume demonstrates some of the ways in which the Internet is precipitating a conceptual shift regarding security.

This chapter combines both of these approaches in order to make four inter-related arguments with respect to satellite imagery and the globalization of trans-parency. First, nonstate actors (firms, IOs and NGOs) are increasingly important as both suppliers and users of satellite imagery. This finding is significant in light of the fact that satellite remote sensing technologies are perhaps the most state-centric information technologies, being firmly rooted in the military and space agencies of the former superpowers. Second, the social and political impact of im-aging satellites can occur through the operation of disciplinary power. In other words, an awareness that one's actions can be observed leads to the internalization of the gaze of the other. Third, the circulation of disciplinary power can open up new possibilities for perceptions of common security and even some elements of collective identity formation—the ironic outcome with respect to space espionage during the Cold War. Whether this is a likely consequence of the global diffusion of commercial spy-quality satellite data is explored in the third section of this paper. Finally, when mapped onto the first argument, the second and third argu-ments suggest that the empowerment of NGOs by information technologies may simultaneously open up new channels for collective identity formation even as it reinforces the circulation of disciplinary power. This line of thinking is explored in the final section of the paper not only with respect to military issues, but also to environmental and humanitarian applications of satellite imagery. Consistent with J. P. Singh's introduction to this book, the diffusion of remote sensing satel-lite technologies has generated important shifts in two dimensions of power: in-strumental and constitutive (or meta-power, as he calls it).

Imaging satellites offer an excellent case for studying some of the sociopoliti-cal dynamics associated with the so-called information revolution. Like a number of other information technologies, most obviously computers and the Internet, imaging satellites were originally developed for military purposes and eventually deployed for civilian purposes by civilian agencies and, more recently, by the pri-vate commercial sector. Contemporary applications include the monitoring of such diverse processes as agricultural productivity, refugee migration and settle-ment, environmental degradation, weapons testing and troop deployments, and the spread of vector-borne diseases. Unlike information technologies which lend themselves to decentralization (i.e., personal computers, fax machines, video cam-eras, and the Internet), remote-sensing satellites offer a less likely case for testing the proposition that information technologies tend to empower nonstate actors. Not only does it require enormous capital investment to launch and operate an imaging satellite, but those states which have traditionally operated spy satellites have been highly protective of their privileged access to "national technical means." This chapter argues that the political history of remote-sensing satellites elucidates both of the ways in which technological change can alter international reality: instrumentally and constitutively.

The foremost political consequence of remote-sensing satellites, transparency, appears to be a straightforward physical effect. Yet, as I argue below, the social meaning of transparency differs widely according to the political context in which satellites operate and imagery is interpreted. Transparency is never absolute for two reasons. First, photo interpretation is as much art as science, and, second, because even when there is consensus on what an image shows, what it means is often highly contentious. While remotely sensed images furnish a certain degree of physical transparency, they cannot render human intentionality and social context transparent. As the experience of the Cold War indicates, the transparency provided by satellites is most likely to move adversaries in the direction of common security arrangements when it is mutual. Recent developments in the diffusion of remote-sensing satellite technologies raise the possibility that this dynamic may be replicated elsewhere and even in other issue areas. Besides the unintended impact upon the relative power and perceived interests of state and nonstate actors, satellite imagery can also facilitate shifts in actors' identities and, consequently, in such purportedly stable features of the international system as anarchy and the security dilemma. Yet new networks of control associated with the diffusion of satellite imagery can still be placed within a larger context of a social episteme rooted in disciplinary power. The ability of nonstate actors to move into what was once the purview of the national security state may not signify a fundamental shift in the nature of power in the international system, even if it does suggest the obsolescence of a state-centric view of world politics. Rather, a host of nonstate actors and new states is being enlisted in the global diffusion of networks of surveillance associated with imaging satellites. This chapter traces the global diffusion of satellite imagery from its early roots in military reconnaissance to the plethora of nonstate uses, ranging from the commercialization of spy data to applications by humanitarian and environmental NGOs.

Foucault concretizes his notion of disciplinary power in his famous discussion of the Panopticon, Jeremy Bentham's design for a model prison. Each inmate is constantly visible from the tower but isolated from other inmates, inducing in the inmate "a state of conscious and permanent visibility that assures the automatic functioning of power" (Foucault 1979, 201). The effects of power are thereby internalized, yielding a form of contingent subjectivity. For a number of reasons, Foucault's work is germane to a sociopolitical understanding of imaging satellites. First, the disciplinary gaze has obvious relevance to a study of remote-sensing technologies, which are inherently oriented towards promoting control through surveillance. Second, Foucault's work shows how technologies of power are not simply exercised by and upon autonomous, preconstituted individuals. Rather, he moves us beyond a purely instrumental conception of power, towards an understanding of how subjectivity itself is shaped by technologies of power. In other words, new technologies can be understood as not only changing the relative power of preexisting social actors, but as helping to constitute those actors by modifying their own self-understandings. Third, disciplinary power, concentrated in the state's administrative apparatus, is emblematic for Foucault of modernity: prisons, the military, schools, insane asylums, and such. Yet, as I argue below, the diffusion of satellite imagery among non-state actors represents a real shift in the institutional locus of disciplinary power.

Ever since Foucault's graphic account of the Panopticon, disciplinary power has taken on a rather nefarious connotation. Yet, as this essay suggests, the disciplinary power associated with remote-sensing satellites can also open up new possibilities for common security arrangements and collective identity formation. The following section argues that this was the somewhat ironic result in the Cold War, during which the primary purpose of remote-sensing satellites shifted from the prevention of surprise nuclear attack to the verification of arms control agreements and the stabilization of nuclear deterrence.

THE BIRTH, LEGITIMATION, AND RECASTING
OF SPACE ESPIONAGE

In the minds of most Americans, the space age and, more specifically, the satellite age began dramatically with the Soviet Union's launch of Sputnik in October 1957 (Oberg 1981). Yet that orbiting ball which aroused such trepidation in the West, carrying only a beeping signal designed to advertise Soviet technological prowess, was not equipped with sensors. The real significance of Sputnik was the large boosters required to launch it, signaling that the Soviets could hurl nuclear warheads across the globe. The United States, not the Soviet Union, was the pioneer in space espionage, and has maintained its position of global leadership in satellite-based remote-sensing into the post-Cold War era.

The satellite age was conceptualized in a 1946 Army Air Force report entitled, "Preliminary Design of a World-Circling Spaceship" (Douglas Aircraft Company 1946). The study predicted that the successful orbiting of a U.S. satellite "would probably produce repercussions in the world comparable to the explosion of the atomic bomb," but the exotic project was shelved in the face of postwar budget cuts (Burrows 1986, 59). Once the Soviets tested atomic and thermonuclear weapons, however, the need for surveillance of the Soviet military machine in order to avert a surprise attack took on a sense of heightened urgency. The closed nature of the Soviet political system, however, presented a major obstacle. In Walter MacDougall's words, "First, and foremost, space was about spying, not because the U.S. was aggressive, but because the USSR was secretive" (MacDougall 1985, 194).

The Killian Report, "Meeting the Threat of Surprise Attack," was submitted to the White House in February of 1955, recommending development of an integrated reconnaissance system which included high-flying aircraft (eventually the U-2 plane) and, most importantly, reconnaissance satellites. Coincidentally, on the same day that this was released, the U.S. National Committee for the International Geophysical Year (IGY) recommended to the National Science Foundation that a scientific satellite be launched as part of the IGY (Day 1996, 240). This convergence of interests led to a top-level policy document known as NSC 5520, calling for the development of a small "scientific" satellite by 1957 as "a test of the principle of 'Freedom of Space.'" (NSC 5520, 6). For if the United States wished to count on the survivability of its spy satellites, as opposed to spy planes which could be shot down over enemy airspace, it needed to establish this principle as an international norm. As M. J. Peterson has compellingly demonstrated, the U.S. accomplished this by successful deployment of analogy, arguing that outer space was more analogous to the open seas than to territorial airspace (Peterson 1997). Thus, from the beginning, remote-sensing satellites required legitimation according to international norms, and therefore some degree of interaction between two rather distinct technological cultures—scientific and military.

While the earliest satellites used television cameras, the prior use of infrared detectors in World War II had disclosed the possibility of using remote sensing to detect different spectral signatures from a given terrain. In other words, because the visible spectrum represents only a tiny range of possible observation, information could also be gathered through the use of spectrometers in the infrared, microwave, x-ray, and gamma ray ranges. The dizzying quantity of potential information made these technologies desirable not just to the military, but to a host of agricultural and forestry experts, geologists, hydrologists, land-use planners, and cartographers. Yet from the beginning, the U.S. military jealously guarded its privileged access to high-resolution remote-sensing satellites.

Meteorological satellites, launched first by the United States in 1960, were under civilian control, yet they also served as legitimators for space espionage by virtue of their apparent innocuousness and their universal utility. In 1960, the Television Infrared Observation Satellite (TIROS-1) began systematic meteorological coverage. The TIROS series later became the National Oceanic and Atmospheric Administration (NOAA) weather satellite system (KPMG 1996, 3). One key element in the United States attempt to legitimize and gain international support for remote-sensing satellites was the Kennedy administration's initiative offering "the free world" open access to meteorological data. TIROS data proved to be an excellent and relatively uncontroversial diplomatic tool. Because its resolution was very low and it primarily monitored cloud cover and weather systems (not land cover), it had little military value. Foreign governments did not object to the imaging of weather systems over their territories since they gained the benefit of free data, thereby adding legitimacy to the norm of "freedom of space."

There has been a common assumption, among policy makers and scholars alike, that greater transparency entails greater international stability. In many cases, this is a valid assumption. Yet, since transparency means that an enemy's targets can be located, satellite data also facilitate a first strike capability. Thus, General Curtis LeMay, head of the Strategic Air Command, believed that the United States should be able to destroy Soviet nuclear forces on the ground, an objective which required huge amounts of detailed photomapping. This more menacing aspect of remote-sensing satellites was not lost on the Soviets. One Soviet official issued the following condemnation of the United States satellite program:

> The main purpose of space espionage is to increase the efficiency of surprise attack, making it possible to knock out enemy missile bases at the very start and thereby avoid a retaliatory blow (cited in Burrows 1986, 135).

Again, the meaning of transparency varies with the social context. Even if the U.S. military command had no intention of launching a surprise attack, the very belief in the minds of Soviet leaders that satellite data might be so used could be destabilizing. Since the flip side of transparency is nakedness, both of the superpowers pursued the development of antisatellite weapons (ASATs).

Nonetheless, as both sides developed pervasive systems for photoreconnaissance, mutual nakedness turned out to be a generally stabilizing force during the Cold War. For one thing, satellites could reveal the falsehood of exaggerated claims, as did the CIA's Corona satellites in the case of the alleged missile gap. Yet the real potential for spy satellites to contribute to stability was not apparent until both sides had them. Indeed, without mutually assured discernibility, the precarious dynamics of mutually assured destruction might easily have ignited a hot war. Perhaps the best example of satellites' ability to contribute to crisis stability came during the Cuban Missile Crisis. Although it was a U-2 spy plane that confirmed the deployment of Soviet missiles in Cuba, satellites also played an important, though not so widely acknowledged, role. First, they gave United States leaders a mountain of background information on the Soviets' overall war-fighting readiness and capabilities, diminishing the likelihood of overreaction. Second, the United States was able to communicate its resolve to the Soviets via the latter's own Cosmos satellites, which could photograph hundreds of readied aircraft bombers en route to the southern United States. Satellites thereby became a crucial element in nuclear deterrence, since they enabled each side to communicate the existence of a credible deterrent force to the other (Klass 1971). Each superpower could thus use the other's spy satellites in order to substantiate its own retaliatory capability. While satellite remote-sensing was born of the need to prevent a surprise attack, the technical ability to render the invisible visible in a social context of mutual distrust provided an unintended cornerstone for mutual deterrence. Today, with the globalization of transparency, it is important to note that stability is more likely to be enhanced when transparency is mutual rather than one-sided. Otherwise, the same sorts of fears as those articulated above by the Soviet general are likely to plague sensed states.

Satellites became part of the architecture of mutual deterrence in another way—through their capacity to contribute to the verification of arms control agreements. Such nuclear arms control treaties as the Limited Test Ban Agreement and the SALT treaties would probably not have been negotiated in the absence of extensive national technical means, as the photoreconnaissance systems came to be known (Office of Technology Assessment 1985). Similarly, more far-reaching arms control proposals, like the popular bilateral nuclear weapons freeze proposal of the 1980s, were premised upon the ability of satellite-based sensors to verify them (Stoertz 1984). Indeed, one of the core principles of arms control, first articulated in SALT I, was that neither side should interfere with the national technical means of the other side. Thus, the exigencies of containing nuclear competition came to legitimate space espionage. More importantly, space espionage was implicated in arrangements bearing some characteristics of a common security regime; even as the superpowers competed fiercely, they were united in their shared aversion to nuclear war. The "nakedness" that was once unacceptable to some U.S. military leaders eventually became not only a way of communicating resolve, but also of making arms control agreements possible. As each side internalized the disciplinary gaze of the other, the superpowers paradoxically fabricated a rudimentary common security regime based upon a mutual interest in preventing nuclear war.

The existence of remote-sensing satellites thereby altered the dynamics of the anarchic international system and the consequent security dilemma by making the actions of each side relatively transparent. The supposed inevitability of a security dilemma in an anarchic world, driving states to arm themselves even when they might prefer not to, is premised upon the inability of states to know their adversaries' actions. The possibility of knowing with a fair degree of accuracy the nature and readiness of an opponent's forces means that one's own actions need not be based upon worst-case scenarios, contrary to the more pessimistic realist readings of the quest for security in an anarchic world. This suggests a new spin on the constructivist claim that the meaning of anarchy changes according to the signaling and intersubjective understandings among the actors. If anarchy is "what states make of it," then technological systems can modify what states make of it (Wendt 1992).

That gaze also entailed institutional arrangements that effectively excluded the citizenry of each of the countries in question. The system was highly secretive and only known in any detail to a small handful of elite officers and political leaders. Yet, as we shall see below, the end of the Cold War opened the door to the diffusion of high-resolution remote sensing across an astonishing range of domains. But even before the end of the Cold War, the basis for that diffusion was being laid by the deployment of two civilian satellite systems: NASA's Landsat in the 1970s and the French SPOT in the 1980s.

NEW ACTORS: CIVILIAN AND COMMERCIAL EARTH OBSERVING SATELLITES

A number of important developments enabled remote-sensing satellites to expand far beyond their original roots in space espionage: 1) the growth of environmental awareness in the 1970s, which highlighted the need for information on land use and environmental problems; 2) the perception that moderate resolution data could appeal to a host of markets; 3) the development of computer-related technologies, especially PC's, GIS, and the Internet, which made satellite data useable in a variety of contexts; 4) the new perception in the 1990s that certain global environmental problems, particularly climate change, can only be understood through the use of large-scale computer modeling based upon enormous quantities of satellite data; and 5) the end of the Cold War, which sent the world's largest military and intelligence establishments in search of alternative missions. The cumulative effect of these factors has been that more and more information has come into the hands of nonstate actors, thereby facilitating the growth of new networks of knowledge and control. As James Rosenau argues in this volume, information technologies, at least as currently deployed, appear to foster a movement away from an hierarchical international system toward the decentralization of power and authority.

The first Landsat acquired low-resolution images (80 meter GSD); Landsat-4 and -5 each generated somewhat better resolution images of 30 meters, but still nothing approaching the accuracy of spy satellites. Nonetheless, their ability to render territory effectively naked was made abundantly clear when the first ERTS images were returned to earth. One image, for instance, revealed a streak of white acid off the coast of New York, indicating that an industrial barge had illegally dumped thousands of gallons of acid iron waste into the Atlantic Ocean only days before (Hall 1992, 52–53). While this incident essentially involved the U.S. "spying" on itself, it dramatically demonstrated the potential for applying satellite-based remote sensing to environmental monitoring and perhaps even industrial espionage.

Although Landsat's resolution was low, its launch sparked international controversy. Countries without access to satellite technology feared that an open-skies policy with respect to civilian remote-sensing satellites would violate their territorial sovereignty. Although they may have harbored such fears earlier regarding military reconnaissance satellites, the fact that superpower images were not available on commercial markets was some source of comfort. NASA's response had the effect of both legitimating civilian remote-sensing satellites and expanding the market for satellite data. First, it argued from international law that there were no legal restrictions on the use of remote-sensing for peaceful purposes. Later, the U.S. took the lead in formulating the United Nations Principles on Remote Sensing, which formalized this principle while granting that sensed states should be given access to data gathered from their territories by civilian satellite systems. (Note, however, that this stipulation does not apply to military and commercial systems.) Second, and most effectively, NASA held out the promise to developing countries that the open dissemination of satellite data would extend, not reduce, their ability to control the development of their resources. To add credence to that promise, NASA established remote-sensing training programs in developing countries and assisted numerous countries in establishing ground stations to receive Landsat data (Lindgren 1988, 34). Within the first decade of Landsat, many countries concluded that transparency and the global diffusion of data were actually in their national interests. By 1980, ten countries had built ground stations and were committed to paying NASA an annual fee of $200,000 for data transmission; dozens more were purchasing Landsat images and data tapes for a host of purposes, including mapping, resource exploration, and environmental monitoring.

By the early 1980s, the primary concern of developing countries was the preservation of open and nondiscriminatory distribution of Landsat data, which they felt was threatened by the Reagan administration's proposal to privatize Landsat (Mack 1990, 188). Many observers believed that Landsat data should remain a public service, analogous to census, cartographic, and meteorological data, and several studies concluded that Landsat could not be successfully commercialized. Despite the objections, the Land Remote Sensing Commercialization Act of 1984 transferred control over Landsat's data to EOSAT, a joint venture of Hughes Aircraft and General Electric, which was later acquired by Lockheed and recently transferred to a Lockheed spinoff, Space Imaging Corporation (Aviation Week and Space Technology 1983, 18). One of EOSAT's first acts, greatly resented by Landsat's user community, was to quadruple the price of Landsat images, causing data sales to drop sharply (Marshall 1989, 24).

The French SPOT, a government-funded commercial system launched in 1986, soon provided an alternative source of satellite data. Because its downlink frequencies were designed to be compatible with Landsat's characteristics, existing Landsat ground stations were able to take advantage of the new source of data (Williamson 1997, 880). The original SPOT returned color images with 20-meter resolution and black-and-white images with 10-meter resolution, nearly crossing the line to resolutions characteristic of military satellites. Since 1978, when President Carter upheld the Pentagon's interests over NASA's by signing a presidential directive 10 meters had been the resolution limit for nonmilitary remote sensing in the United States (Zimmerman 1988, 48). SPOT not only became Landsat's first competitor, but also demonstrated the technological creep towards the globalization of ever higher resolution imagery. While SPOT's top customer, ironically, has become the U.S. military, markets for SPOT imagery include agricultural and forestry applications, mining and petroleum geology, hydrology, and land-use planning. With the launch of SPOT, the United States no longer enjoyed a monopoly on commercial Earth-scanning satellites. Yet, despite concerns by Landsat supporters that SPOT's data sales would undercut Landsat, SPOT on the contrary expanded the overall imagery market by demonstrating new data applications (Williamson 1997, 880).[1]

As the Cold War drew to a close, the Soviets decided that their satellites' commercial value outweighed their espionage value. In 1989, a Soviet system using conventional camera film began marketing images with resolution between 2 and 5 meters, substantially lowering the threshold between military and civilian space data.[2] However, they cannot be properly called spy data because of the time lag in retrieving them (Spector 1989, 16). United States policy under the Reagan and Bush administrations, which put licensing authority in the hands of the Departments of Defense and State, effectively obstructed the entry of U.S. commercial satellite operators with high-resolution products into the global marketplace. Consequently, the Clinton administration amended the Reagan rule so that licensing applications for remote-sensing satellite systems with capabilities already available or in the planning stages would be favorably considered (The White House 1994, 243–244).[3]

Soon thereafter, U.S. military and intelligence firms concerned about the future of contracts in the post-Cold War era and encouraged by the declining cost of launching imaging satellites, entered the marketplace. Despite the enormous investment, ranging in the hundreds of millions of dollars, entailed in building and launching high-resolution satellites, these firms' optimism was fueled by the growth of a supportive technological infrastructure: inexpensive personal computers and software packages powerful enough to process satellite images, CD-ROM disks capable of storing massive quantities of digital data, and widespread Internet access for searching and transferring large data files (Williamson 1997, 883). Beginning in 1993, a number of companies obtained licenses to operate high-resolution satellite systems, including Orbital Imaging, Worldview (later

Earthwatch), and Space Imaging. The first of the new generation of commercial high-resolution satellites, Space Imaging's IKONOS, began returning images in 1999. Within the next five years, approximately twenty commercial satellites capable of returning images with better than ten-meter resolution are due to be launched—the majority by U.S. firms. Thus, high-resolution satellite imagery is migrating from the "deep black" of military espionage into a host of civilian and commercial applications—including the agricultural sector, mineral exploration, the real estate industry, municipal utilities, and environmental monitoring.

Before examining these new networks of control, however, let us look at the one ready-made market for spy-quality satellite data—the militaries of states that do not possess space espionage capabilities. The argument was made above that the disciplinary gaze of spy satellites during the Cold War had the unforeseen effect of drawing the superpowers into some elements of a common security arrangement premised upon mutual transparency and a shared aversion to nuclear war. Is open access to commercial high-resolution satellite imagery likely to have an analogous impact on regional tensions around the world?

SPY IMAGES EVERYWHERE

Within a few years, the commercialization of high-resolution satellite data could foster mutual transparency in situations of potential conflict around the world, particularly in instances where only one state has an indigenous remote sensing capability. More interesting in terms of agency in world politics would be the extension of the disciplinary gaze of space espionage to international public opinion and domestic civil society, suggesting the possibility of involving NGOs in a game that heretofore was played almost solely by national governments.

Consider, for instance, the nuclear weapons test by India in May 1998 in a context in which India had an indigenous remote-sensing capability while Pakistan did not. In fact, India's IRS-1C satellite, with a resolution of less than six meters, was returning some of the highest resolution images then available on commercial markets. The U.S. intelligence community, with access to satellite images with resolutions of perhaps six inches and upwards of two thousand photo analysts at its disposal, was apparently taken by surprise by India's actions, perhaps because monitoring the Indian test site was not a top priority for the U.S. While Pakistan could have purchased low and moderate resolution on commercial markets, these may not have been sufficient to confirm India's preparations for nuclear weapons testing. But what if a dozen commercial high-resolution satellites had been in orbit, as will most likely be the case in five years? Monitoring the Indian test site would certainly have been a high priority for the Pakistani government. Had Pakistan detected preparations for a nuclear test, it may not necessarily have been able to prevent India from following through with its plans, but it could have brought the evidence to the international community and perhaps even used it to spark domestic debate within India itself. Indeed, users of commercial satellite data would not be nearly so limited in publicizing images as users of national intelligence data because the veil of secrecy surrounding sources and methods for national technical means would not exist for commercial satellites. On the contrary, commercial satellite operators would probably be glad for the publicity.

Of course, any state intent upon pursuing a specific course of action could resort to what intelligence officers call CCD: camouflage, concealment and deception. As Foucault observes, wherever power operates, there is resistance. Activities can be concealed with nets, canvas, roofs or vegetation, or they might be performed at night or under cloud cover. The point, however, is that the proliferation of observation satellites passing over at various times decreases the likelihood that such efforts at resistance will succeed. Moreover, some of those satellites may be equipped with Synthetic Aperture Radar (SAR) sensors capable of imaging at night or under cloud cover. For instance, Canada's RADARSAT, launched in 1995, is equipped with such a sensor. While RADARSAT currently generates images with only 8-meter resolution, Canada hopes to launch a new radar satellite with a resolution of three to five meters. Transparency may never be absolute, but as it increases, the possibilities for concealment are narrowed.

Up to this point, this paper has only discussed the impact of satellite imagery on international conflict in the context of bilateral relations. Yet, given the nature of international conflict since the end of the Cold War, it is worth looking at the potential impact of satellite imagery on a multilateral conflict situation. Indeed, a key factor in U.S. policy decisions to return the Landsat system to government operation and to license U.S. companies to launch high-resolution satellites in the 1990s was the utility of Landsat and SPOT data during the 1991 Persian Gulf Conflict. Despite both the costs of building Landsat 7 during a period of budget tightening and the dangers inherent in allowing commercial firms to build the equivalent of spy satellites, U.S. officials felt that the alternative of leaving the market to foreign firms was worse (Williamson 1997, 881; Gordon 1993). Prior to and during Operation Desert Storm, maps created from Landsat and SPOT data had one clear advantage over images generated by reconnaissance satellites: they could be shared openly with allies. Since the military value of high-resolution data will be substantially greater, the benefits of openness in multilateral conflict situations will be that much greater. If, as some have argued, multilateralism is the wave of the future in international relations—whether in situations of conflict or cooperation—then the overall utility of commercial and civilian satellites is likely to exceed that of national spy satellites simply because the range of situations in which the latter can be used is limited. With spy-quality images available on commercial markets in the coming years, many states will be able to partake in the disciplinary gaze of remote-sensing satellites for a range of purposes, including treaty verification, U.N. peacekeeping operations, threat assessment, signaling, or (more ominously, and in conjunction with GPS) targeting.

The possibility that commercial satellite data could be used for offensive purposes by an adversary was another factor driving the U.S. decision to license high-resolution imagery vendors. American companies, unlike foreign ones, can be controlled by the U.S. government during a crisis. For example, had Iraq been able to acquire Landsat and SPOT images prior to the allied invasion, it could have detected the famous "Left Hook" strategy in which allied forces massed in Saudi Arabia rather than attacking solely by sea. Purchasing only two SPOT scenes of the area some time later, photo analyst Vipin Gupta has demonstrated that the Left Hook maneuver was easily detectable using commercial moderate-resolution images (Gupta 1998). Although SPOT was effectively requisitioned by the U.S. government (in consultation with France) during the Gulf conflict, such cooperation is not always guaranteed. Thus, in the face of the globalization of transparency, states may wish to exercise control over commercial satellite operators and, indeed, they still retain a substantial ability to do so.

While commercialization entails the migration of satellites' disciplinary gaze beyond the exclusive purview of the state, the critical functions of licensing and regulation remain in the hands of the state. The government will retain at a minimum the right to stop transmission of data during wartime. Yet satellite operators fear that less urgent foreign policy and national security considerations could lead to burdensome restrictions on their ability to sell their wares. The 1996 Bingaman-Kyl Amendment to the National Defense Authorization Act, prohibiting U.S. firms from collecting or selling high-resolution satellite data of Israeli territory, fueled their concerns and highlighted the extent to which the state—or at least one particular state—continues to exert its influence in this arena (Baker 1997, 8). As Jonathan Aronson argues earlier in this volume, governmental regulations have a substantial impact upon what satellite operators can and cannot do.[4] The diffusion of imaging technologies may foster decentralization and the proliferation of networks beyond the state, but the state is not thereby rendered powerless or anachronistic.

Nonetheless, the profusion of civilian and commercial satellite imagery opens up new opportunities for nonstate actors to be involved in traditional national security issues. Even at $5,000 per scene (the price of one RADARSAT image), the cost of purchasing imagery is small compared to the cost of launching a satellite. A much smaller, but politically significant, market would be nongovernmental organizations (NGOs) and the media. A handful of NGOs has used satellite images to advocate for peace and disarmament or to assist in the verification of arms control agreements. The U.K.-based VERTIC, Verification Technology and Information Center, has used imagery to compile its yearly Verification Report on arms control and environmental treaty compliance. Vipin Gupta, who worked with VERTIC on some of his early photo interpretation projects, has demonstrated the utility of publicly accessible satellite imagery in monitoring the nuclear activities of China, India, and Algeria (Gupta 1992; Gupta and McNab 1993; Gupta 1995; Gupta 1997). The U.S.-based "Public Eye Project," sponsored by the Federation of American Scientists, has posted declassified images from the CIA's Corona satellites on the Internet. John Pike, director of the project, hopes to obtain funding to purchase new commercial high-resolution images that will enable citizens to monitor military developments around the world (See http://www.fas.org). While the incorporation of the disciplinary gaze of imaging satellites into domestic and global civil society is admittedly embryonic, these instances reveal the potential for commercial high-resolution satellites to open up the relatively closed world of national security politics to the scrutiny of nonstate actors. Although few NGOs possess the resources necessary to purchase and utilize satellite imagery, the possibility remains that the combination of a plethora of competing commercial satellites along with the availability of cheap PC-based software for manipulating images could bring down the cost sufficiently for them do so.

Another set of nonstate actors, however, is less likely to be deterred by cost from using pictures from orbit to bring military issues before the public—the news media. United States news organizations have used satellite imagery since 1986 when ABC, followed by NBC, CBS and The New York Times, used SPOT photos to report on the Chernobyl nuclear disaster (Blumberg 1991). The potentially dramatic impact of the media's use of commercial high-resolution images was demonstrated prior to Operation Desert Storm, when ABC purchased two two-meter resolution images from the Soviet commercial satellite company, Soyuz-Karta, for $1560 apiece. Photo analysts working for ABC News could find no trace of the 265,000 Iraqi troops which U.S. officials claimed were massed at the Saudi-Kuwait border in the fall of 1990. Nonetheless, ABC decided not to run the story because their photos did not include a section of southern Kuwait where the troops may have been. The St. Petersburg Times soon purchased the third photo; their analysts also found no evidence of a massive Iraqi presence. The Pentagon insisted that the troops were there, but would provide no visual evidence, declaring, "We'd like it to remain a mystery to Saddam what our intelligence capabilities are. We are not going to make our intelligence public" (Heller 1991).

Because of its timing, this news story had little political impact. By the time the story broke, less than two weeks before the U.S.-led invasion, the momentum was so great that there was little likelihood of revisiting the public debate of the preceding months. Nonetheless, the incident indicates that commercial high-resolution satellite imagery could enable the news media to play a much more active role in monitoring and publicizing military developments.[5] A commercial market of spy-quality data would at least partially lift the veil of secrecy behind which states have traditionally conducted their military operations, giving citizens access to the disciplinary gaze of orbiting satellites.

NONSTATE ACTORS AND NEW NETWORKS OF SURVEILLANCE

The diffusion of satellite imagery, coupled with the advent of desktop computer programs capable of manipulating the large databases associated with satellite imagery and transmitting them over the Internet, is creating new networks of surveillance involving NGOs working on nonmilitary issues as well. Satellites can enable citizens, mostly in industrialized countries, but to a growing extent in developing countries as well, to monitor such diverse phenomena as crop and weather conditions, deforestation, marine habitat, and the movement of large numbers of refugees. By serving as a powerful tool of legitimation, Earth observations from space can offer visual evidence to support NGOs' positions.

The empowerment of NGOs by remote-sensing technologies simultaneously erodes states' abilities to control information about developments in their own territories even as it intensifies the circulation of disciplinary power. The slippage of information beyond the state does not mean that NGOs and states must stand in an adversarial relationship to one another. In some cases they do; in others they do not. But it does mean that states' ability to control the flow of information about their own activities and within their own territorial borders is being eroded. Furthermore, the disciplinary power of satellites in nonmilitary matters tends not to involve adversaries internalizing the gaze of the other, but actors entwined in networks of power within a particular domain, for instance, climate change, deforestation or humanitarian relief. This section looks at the use of satellite imagery as a technology of power by environmental NGOs (ENGOs) and humanitarian NGOs (HNGOs).

ENGOs have deployed satellite data as a powerful tool of legitimation. The Nature Conservancy, for instance, has used satellite data to evaluate biodiversity and assess the health of biotic communities in their efforts to monitor enforcement of the U.S. Endangered Species Act (Stein 1996). The Britain-based Coral Cay Conservation uses Landsat TM data along with aerial photography in order to document reef destruction around the world and to generate GIS coral reef habitat maps which are used in ecosystem management initiatives (Harborne, personal communication, 6 May 1998). The popular World Resources Institute annual reports base their estimates of global and regional deforestation primarily on satellite data (World Resources Institute 1998). A recent WRI report combined satellite data and fossil records to conclude that only 20 percent of the planet's original ecologically intact forests remains today (Bryant et al. 1997). In all of these instances, satellite data is used to authenticate findings for an audience that might include policy makers, the public, or potential donors.

Perhaps the greatest asset of satellite imagery is its visual character. As one individual working to stop logging in British Columbia's Clayoqot Sound, the last remaining expanse of temperate rainforest, declared, "Satellite images are totally convincing. You show people a map, and they can see clearly what's left" (quoted in Clayton 1996, 5). Satellite imagery can make a distant, abstract ecological crisis seem present and tangible. Consider, for instance, the widely circulated video clip of the Antarctic ozone hole during the negotiations for the 1987 Montreal Protocol. That video, a composite of satellite imagery which was broadcast to treaty negotiators, the U.S. Congress and television audiences all over the world, transformed an utterly invisible event into a dramatic global ecological crisis. Thus, the Antarctic ozone hole, which the negotiators agreed to ignore because its causes were unknown, nonetheless had a profound impact on the negotiations (Litfin 1994).

With the advent of commercial high-resolution imaging satellites, new information technologies that make satellite data accessible in a timely manner, and a heightened involvement of the military and intelligence establishment in humanitarian crises, the stage has been set for HNGOs to incorporate satellite imagery into their work. HNGOs have a good deal of experience in bringing seemingly remote crises into the living rooms of television audiences around the world, but they often find themselves in need of a visual source of objective information. Satellite imagery can help them to answer such questions as how many people are in need of assistance, precisely where they are and which way they are going, and what local geographic and resource factors are likely to affect assistance efforts. Not only can satellite imagery help HNGOs in formulating their own relief operations, it can help them to communicate the gravity and scope of a problem to policy makers, the public, and donors. As more spatially focused and temporally compressed imagery becomes available with the commercialization of high-resolution imaging satellites, HNGOs are likely to find Earth observations a useful tool for legitimation.

As Martha Finnemore has demonstrated (1996), normative understandings about which human beings merit international intervention have changed dramatically in this century. In past centuries, norms of humanitarian intervention applied only to whites and Christians. By the late twentieth century, a combination of decolonization and a globalized telecommunications system has facilitated what might be called the universalization of humanity, such that famines, natural disasters, and genocide anywhere on the planet have become cause for international concern and action. The global reach of Earth observations from space is not only compatible with this normative trend; it can also help reinforce it in some very practical ways. Terrain mapping can assist in locating areas with appropriate topography, land cover, and access to water for the settlement and repatriation of refugees. Likewise, satellite data can help determine the environmental impact of large-scale movements of refugees for purposes of reforestation and habitat conservation. In areas with extensive cloud cover, especially in the tropical cloud belt, radar images can be useful in locating and tracking refugees; satellite-based observations of campfires can provide important data for estimating total numbers of refugees. For rapid-onset disasters like floods, cyclones, earthquakes, and volcanic eruptions, satellite imagery can be coupled with on-site GIS/GPS capacity to produce maps with key hazards and infrastructural damage pinpointed accurately. In programs using food aid, imagery can aid in tracking the movement of food commodities along supply chains as well as in pinpointing likely obstacles and bottlenecks, even in closed societies like North Korea. Thus, the global reach of satellite imagery is consistent with the universalization of humanity and the internationalization of humanitarian assistance.

Nonetheless, the experience of ENGOs suggests an element of caution. First, even at one-meter resolution, imaging satellites cannot discern the social causes of either ecological or humanitarian crises. Because human agency all but vanishes from the perspective of outer space, crises may be mistakenly reduced to physical processes and root social causes thereby ignored. For humanitarian relief efforts concerned with short-term amelioration, this may not be a huge problem. But if either prevention or a deeper understanding of the causal roots of the problem is the objective, satellite imagery is of limited utility. In any case, both ENGOs and HNGOs will find that satellite data needs to be supplemented with substantial "ground truthing." Satellite images cannot reveal that logging trucks in Malaysia are owned by Mitsubishi. Nor can they reveal that people in a drought-ridden part of Africa are not starving because men from those villages work elsewhere and send their wages home. Thus, there is a strong need to pair satellite data with sociological and anthropological appraisal tools on the ground.

Moreover, as ENGOs in developing countries have pointed out, satellite data is ahistorical in the sense that it shows current, but not past, levels of ecological degradation. Landsat and SPOT images of deforestation in the tropics, for instance, not only render human agency invisible, they also render invisible previous centuries' deforestation in Europe and North America, thereby tending to reinforce the perception that deforestation is caused by developing countries. Nor can satellite images reveal that most of tropical timber exported from Southeast Asia is consumed by people in industrialized countries. Similarly, while Earth observations from space are likely to prove useful to HNGOs in future relief operations, they cannot, by their very nature, uncover the deeper political and economic causes that are often at the root of "natural" disasters. While satellite data can reveal much about the earth's surface, social meanings (and hence policy implications) are not rendered transparent so easily.

The fact that satellite data renders human agency invisible is related to a second caveat, which is that while having good information is preferable to having poor information, it does not necessarily generate good policy. ENGOs can cite numerous examples where further research has been a substitute, rather than the basis, for sound policy. One can imagine a crisis in which precise numbers and geolocations of starving refugees are known, along with digital elevation maps of their camps, but little is done to alleviate their suffering. There is no simple correlation between having detailed information about a problem and knowing how to respond. If the habitat of an endangered species is being destroyed on private land, should the federal government establish a protected wilderness area, or should it offer tax incentives and zoning provisions to induce the owners to preserve the ecosystem? If imagery reveals 500,000 refugees in a zone of conflict, should the United States or the U.N. send troops (and, if so, on what sort of mission?) or should HNGOs and intergovernmental organizations assume full responsibility for the relief effort?[6] Accurate information may be an important factor in formulating effective policy responses, but it is by no means a panacea.

While satellite imagery provides a powerful legitimating tool, it can overshadow or even displace other forms of information. The use of imagery and GIS packages by indigenous groups for mapping their customary land rights and documenting the role of governments and multinational corporations in ecological destruction, for instance, requires that these groups legitimate their positions on the basis of high technology rather than more traditional forms of knowledge. Similarly, international ENGOs based in Europe and North America have found that using satellite data for conservation purposes in the tropics can have the unintended consequence of undercutting local aerial photography companies, an ironic consequence if economic development is one of their objectives. If knowledge is power, then deploying satellite images and the associated digital tools is a question of political economy, and not just a technological issue. As Comor argues later in this volume, information technologies are thoroughly embedded in the global economy; thus, remote sensing satellites, deployed and utilized primarily in the industrialized countries, are not divorced from economic power—especially in the case of commercial satellites.

For both HNGOs and ENGOs, the use of Earth observations from space, while often an effective tool, is also costly. A single image, in itself relatively useless in the absence of interpretive techniques, can cost anywhere from $500 to $5000. Very few NGOs have the technically trained staff needed to acquire, interpret and manipulate satellite images. For instance, the U.S.-based Wilderness Society is one of the only ENGOs with an in-house remote sensing facility, acquired as a donation, but only after substantial fundraising could the group afford to hire the technical staff needed to operate it (interview with Janice Thomson, Remote Sensing Coordinator for the Wilderness Society). Because of the cost and high level of technical expertise involved in using satellite imagery, ENGOs have found that they can only make effective use of the technology by engaging in a whole range of partnerships with individuals and organizations that contribute to their work. Networks of surveillance that include NGOs are therefore likely to involve ties to universities and research institutions, relationships with state agencies and international organizations, corporate sponsorship, and perhaps even links to the military/intelligence community.

Scientific studies on a host of environmental problems, ranging from biodiversity conservation to ozone depletion to declining fisheries to deforestation, have employed satellite imagery, and ENGOs have become adept at piggybacking their own work on existing studies. While HNGO ties to the scientific community are not as strong as those of their environmental counterparts, there is some evidence that those ties are increasing. In the first project of its kind, researchers at the Nansen Environmental and Remote Sensing Center in Norway have begun to develop near real-time image processing and interpretation techniques for using high-resolution satellite images to support humanitarian relief operations (Bjorgo 1996). The Center for PVO-University Collaboration was founded on the premise that the work of PVOs (private voluntary organizations, another term for NGOs) is greatly enhanced by ties to university researchers. Most of the two dozen international NGOs associated with the center work on humanitarian and environmental issues, and many of them have used or plan to use satellite imagery in their work (See http://www.wcu.edu/UnivPVO).

Interestingly, the end of the Cold War has led the intelligence community to become involved with both humanitarian and environmental issues. The need for institutional legitimation in the vacuum left by the Cold War, along with a heightened sense of environmental awareness, has no doubt contributed to the U.S. intelligence community's newfound interest in global ecology. On humanitarian issues, the intelligence community has been brought in by the multilateral nature of the problems and the U.S. military's need to coordinate with multiple actors, including HNGOs. While imagery from the U.S. intelligence community's state-of-the-art satellites would be of great utility to both HNGOs and ENGOs, the key obstacle, of course, is secrecy. To make images available would be to divulge the technical capabilities of satellites whose names are not even public knowledge, thereby, according to the intelligence community, enabling potential adversaries to more easily conceal their activities. Thus far, the intelligence community's environmental activities (the MEDEA project and the Global Fiducial Data Program, for instance) have had no discernible impact on the work of ENGOs (Thomas 1997).

In its support to disaster and humanitarian relief operations, however, the intelligence community may have a greater incentive to share imagery when U.S. troops are involved and the lives of Americans are at stake. Military logistics in such situations are likely to involve coordinating activities with HNGOs and even protecting their supply lines and other operations. In many cases, as in Somalia, Bosnia, and Rwanda, HNGOs operate on the ground long before military personnel arrive on the scene. Consequently, the intelligence community is coming under increasing pressure to share information, including satellite imagery, with HNGOs. Indeed, one of the key findings of a 1995 CIA-sponsored study is that intelligence for humanitarian relief purposes too often "excludes important players, and thereby limit[s] the value of the information provided" (Constantine 1995, 1).

A multitude of social, environmental and economic problems have proven themselves too intractable for states alone to address. NGOs have proliferated to fill the void. Simultaneously, political developments and the information revolution have enabled NGOs to participate in technologies of surveillance previously confined to the military arena. User communities, primarily scientific and government, were initially stimulated by Landsat and later by SPOT. The multibillion dollar GIS industry and Internet accessibility have helped to spread satellite imagery far and wide, with the commercialization of high-resolution satellite data contributing to that momentum. High-resolution satellite imagery, until recently monopolized by the national security agencies of the superpowers, is now freely available to anyone with access to a credit card and the Internet.

CONCLUSIONS

Within the last decade, the disciplinary gaze of satellites which once helped to normalize nuclear deterrence between the superpowers has been turned towards a plethora of military, economic, social, and environmental purposes involving ever more expansive networks of surveillance. Technologies of surveillance, from population census to welfare rolls to weather prediction, have been the basis for the state's administrative power throughout the modern era. Indeed, 'statistics' and 'state' are derived from the same Latin root (to stand); not coincidentally, the large-scale collection of statistics began with the emergence of the modern state (Taylor and Johnston 1995). The early satellite era during the Cold War is a prime example of this concentration of technologies of surveillance in the bureaucratic apparatus of the state. Space espionage, providing a safe and unobtrusive way for the superpowers to monitor military developments, made both sides conscious that their deployments were largely visible to the other. The circulation of disciplinary power had the ironic effect of undercutting the more pernicious dynamics of the security dilemma by bringing the two arch-rivals into elements of a common security regime. The commercial availability of high-resolution imagery may lead to a similar outcome in other conflict situations, with the important exception that the absence of a perceived need to protect sources and methods will permit domestic and transnational civil society to share in the global gaze.

With civilian and commercial imaging satellites, new networks of surveillance have sprung up around a host of issues and have drawn in an increasingly diverse array of state and nonstate actors. This general trend lends credence to Rosenau's claim that the diffusion of information technologies is undercutting the ability of states to exercise control and authority. The end of the superpower monopoly on space espionage, however, did not disempower the state, but rather brought previously excluded states into the game. States still enjoy a significant degree of control even in the face of commercialization because of their ability to license commercial satellites, to outlaw the export of turnkey systems, to exercise shutter control, to compete in markets, to mandate data purchasing policies for their agencies, and (in the most far-fetched scenario) to shoot the satellites down. Thus, while the general trend is clearly in the direction of the diffusion of satellite surveillance beyond the state, the case should not be overstated. New networks of surveillance may decenter the state, but they do not render it obsolete.

What, then, is the impact of these shifts on the self-understandings of social actors and the nature of power in world politics? With access to satellite imagery, those who were excluded from past policy debates on military, economic, social, and environmental questions may now be able to participate. More fundamentally, the disciplinary gaze of space espionage is no longer fixed in the state, but is increasingly dispersed across multiple social and political levels. As Ronald Deibert has argued, information technologies may undermine rather than enforce the notion of the modern state drawn from Foucault's model of disciplinary power. Rather than investing a single privileged center with a panoptic gaze, information technologies seem to facilitate the growth of a "dispersed surveillance web" (Deibert 1997, 166–171). The diffusion of satellite imagery appears to be consistent with that trend, lending support to John Ruggie's suggestion that the unitary perspective of state authority may be on the verge of being displaced by multiperspectival sources of governance (1993).

Before seizing on any grandiose conclusions about a transition from modernity to postmodernity, we must also note that the diffuse surveillance web associated with the dissemination of satellite imagery has, in effect, enlisted a host of nonstate actors in the larger social project of prediction and control associated with modernity. Imaging satellites function simultaneously as symptom, expression, and reinforcement of modernity's dream of knowledge as power. The adversary's military disposition, agricultural production, marine fisheries, global climate change, land use patterns, movements of refugees: all are to be known, predicted, and normalized. In some cases, the control that derives from the disciplinary gaze of satellites has favorable consequences, as it seems to have during the Cold War. Nonetheless, the diffusion of satellite imagery is contributing simultaneously to the decentralization of surveillance and to its universalization.

NOTES

I am grateful to J.P. Singh and Stephen McDowell for their detailed and extremely helpful comments on an earlier draft of this paper.

1. For instance, the European Union, turning its disciplinary gaze on its agricultural producers, paid for the vegetation monitor on the recently launched SPOT 4 in order to verify that farmers are actually growing what they claim to be growing.

2. Until the 1999 launch of Space Imaging's IKONOS satellite, the Russian SPIN-2 product, with a two-meter resolution, was the highest resolution imagery commercially available. Designers of the Terra-Server, a joint effort by Microsoft Corporation, Aerial Images, Sovinformsputnik, and Digital Equipment Corporation, promise that their new online world atlas will guarantee that "we can see ourselves as we really are." SPIN-2 images of major cities can be purchased through the Internet for $25. See http://www.spin-2.com.

3. For a detailed analysis of the Clinton administration's policy decision, see Baker 1997.

4. Of course, some states could simply prevent satellites from collecting and transmitting imagery by shooting them down. The costs of doing so, however, would be enormous, both materially in terms of the orbital debris that could disrupt telecommunications satellite transmissions everywhere and politically in terms of committing an unprecedented act of war in outer space. The fact that states do not shoot down satellites points to the resilience of the international norm of freedom of space, despite the inclusion of such scenarios in U.S. information warfare strategic planning.

5. We should note the danger, however, that news organizations, in their rush to break an important story or because of their lack of familiarity with photo-interpretation, could be wrong. For instance, following India's nuclear weapons test in May 1998, *Newsweek* included a satellite image that mistakenly identified the Indian nuclear test site (Vipin Gupta, personal communication, May 21, 1998).

6. This, in fact, was the situation during the massive refugee migration in Central Africa's Great Lakes region from 1996 to 1997. While the U.S. intelligence community had access to high-resolution images, it refused to share them with HNGOs, most of which were pressing for U.S. or U.N. involvement at the time. Had commercial high-resolution satellite data been available at the time, the HNGOs would have had access to an alternative source of information.

REFERENCES

Aviation Week and Space Technology. 1983. Report Criticizes Landsat Commercialization. (May 9) 118: 18.

Baker, John C. 1997. Trading Away Security: The Clinton Administration's 1994 Decision on Satellite Imaging Exports. *Pew Case Studies in International Affairs, Case 222*. Washington, D.C.: Institute for the Study of Diplomacy.

Bjorgo, Einar. 1996. RefMon: Refugee Monitoring Using High-resolution Satellite Images. *Nansen Environmental and Remote Sensing Center.* Bergen, Norway. (http://www.nrsc.no:8001/~einar/UN/remon_abstract.html).

Blumberg, Peter. 1991. Satellite Picture Puzzle: No Iraqis. *Washington Journalism Review.* (May) 14: 13–14.

Bryant, Dirk, Daniel Nielsen, and Laura Tangley. 1997. *The Last Frontier Forests: Ecosystems and Economies on the Edge.* Washington, D.C.: World Resources Institute.

Burrows, William E. 1986. *Deep Black: Space Espionage and National Security.* New York: Random House.

Center for PVO-University Collaboration website, http://www.wcu.edu/UnivPVO.

Clayton, Mark. 1997. Got an Earthly Cause? *Christian Science Monitor.* (May 8) 1.

Constantine, G. Ted. 1995. Intelligence Support to Humanitarian-Disaster Relief Operations. *Center for the Study of Intelligence Monograph.* Langley, VA: Central Intelligence Agency.

Day, Dwayne. 1996. Invitation to Struggle: The History of Civilian-Military Relations in Space." In John M. Logsdon, ed. *Exploring the Unknown: Selected Documents in the History of the U.S. Civilian Space Program, Vol. 2: External Relationships.* Washington, D.C.: NASA.

Deibert, Robert. 1997. *Parchment, Printing, and Hypermedia: Communication in World Order Transformation.* New York: Columbia University Press.

Douglas Aircraft Company. 1946. *Preliminary Design of an Experimental World-Circling Spaceship.* Report No. SM-11827.

Federation of American Scientists website, http://www.fas.org.

Finnemore, Martha. 1996. Constructing Norms of Humanitarian Intervention. In Peter J. Katzenstein, ed. *The Culture of National Security: Norms and Identity in World Politics.* New York: Columbia University Press. 153–185.

Foley, Theresa M. 1988. Pentagon, State Department Granted Veto Over U.S. Remote Sensing Satellites. *Aviation Week and Space Technology.* (July 20) 20–22.

Foucault, Michel. 1979. *Discipline and Punish: The Birth of the Prison.* New York: Vintage.

Gupta, Vipin. 1992. Algeria's Nuclear Ambitions. *International Defense Review.* 4: 329–331.

——. 1995. Locating Nuclear Explosions at the Chinese Test Site near Lop Nor. *Science and Global Security.* 5: 204–244.

——. 1997. "Investigating the Allegations of Indian Nuclear Test Preparations in the Rajasthan Desert, *Science and Global Security.* 6: 101–188.

——. 1998. Detecting Troop Concentrations Using SPOT Data: The 1991 Persian Gulf War. Presentation for *Secret No More: The Security Implications of Global Trans-*

parency. Workshop sponsored by the National Air and Space Museum, Washington, D.C. (May 21–22)

Gupta, Vipin, and Philip McNab. 1993. Sleuthing From Home. *The Bulletin of the Atomic Scientists*. 49 (10): 44–47.

Hall, Stephen. 1992. *Mapping the Next Millennium: The Discovery of New Geographies*. New York: Random House.

Heller, Jean. 1991. Photos Don't Show Buildup, *St. Petersburg Times*. (Jan. 6) A-1.

Klass, Philip J. 1971. *Secret Sentries in Space*. New York: Random House.

KPMG. 1996. *The Satellite Remote Sensing Industry: A Global Review*. Washington, D.C.: KPMG.

Lindgren, David T. 1988. Commercial Satellites Open Skies. *Bulletin of Atomic Scientists*. (April) 30–36.

Litfin, Karen T. 1994. *Ozone Discourses: Science and Politics in Global Environmental Cooperation*. New York: Columbia University Press.

MacDougall, Walter. 1985. *The Heavens and the Earth*. New York: Basic Books.

Mack, Pamela. 1990. *Viewing the Earth: The Social Construction of Landsat Satellite System*. Cambridge, MA: The MIT Press.

Marshall, Eliot. 1989. Landsat: Drifting toward Oblivion? *Science*. (February 24) 243: 24.

Meyer, Stephen M. 1984. Verification and Risk in Arms Control. *International Security*. 9.

National Security Council. 1955. Draft Statement of Policy on U.S. Scientific Satellite Program. *NSC 5520*. Washington, D.C.: National Archives and Records Administration.

Oberg, James. 1981. *Red Star in Orbit*. New York: Random House.

Office of Technology Assessment. 1985. *Anti-Satellite Weapons, Countermeasures, and Arms Control*. Washington, D.C.: Government Printing Office.

Peterson, M. J. 1997. The Use of Analogies in Developing Outer Space Law, *International Organization*. 51(2): 245–74.

Rosenau, James N. 1990. *Turbulence in World Politics: A Theory of Change and Continuity*. Princeton: Princeton University Press.

Ruggie, John G. 1993. Territoriality and Beyond: Problematizing Modernity in International Relations. *International Organization*. 46(1): 139–174.

SPIN-2 website, http://www.spin-2.com.

Stein, Bruce. 1996. Putting Nature on the Map. *Nature Conservancy*. (January/February) 24–27.

Spector, Leonard. 1989. Keep the Skies Open. *Bulletin of Atomic Scientists*. (September) 14–20.

Taylor, Peter J., and Ronald J. Johnston. 1995. Geographical Information Systems and Geography. In John Pickles, ed. *Ground Truth: The Social Implications of Geographic Information Systems.* New York: The Guilford Press.

Thomas, Gerald B. 1997. U.S. Environmental Security Policy: Broad Concern or Narrow Interests? *Journal of Environment and Development.* (December) 6(4): 397–425.

Wendt, Alexander. 1992. Anarchy is What States Make of It. *International Organization.* 46: 391–425.

The White House, Office of the Press Secretary. 1994. Foreign Access to Remote Sensing Capabilities. *Space Policy.* (August) 10(3): 243–244.

Williamson, Ray A. 1997. The Landsat Legacy: Remote Sensing and the Development of Commercial Remote Sensing. *Photogrammetric Engineering and Remote Sensing.* (July) 63(7): 877–885.

World Resources Institute. 1998. *World Resources 1998.* Washington, D.C.: World Resources Institute.

Zimmerman, Peter D. 1988. Photos from Space: Why Restrictions Won't Work, *Technology Review.* (May) 91.

CHAPTER FOUR

INFORMATIONAL META-TECHNOLOGIES, INTERNATIONAL RELATIONS, AND GENETIC POWER: THE CASE OF BIOTECHNOLOGIES

SANDRA BRAMAN

INTRODUCTION

Meta-technologies are politically critical because they vastly expand the capacity of state and non-state actors to exercise genetic power—control over the informational bases of the materials, structures, and ideas that are the stuff of power in its instrumental, structural, and symbolic forms. The increasing use of meta-technologies to exercise genetic power is changing the rules of the game in international relations by disrupting long-standing structures, changing the relative weights among classes of players, and turning attention from products to processes. Meta-technologies are always informational; as a class they include those that process biological information, biotechnologies, as well as those that process digital information, or digital information technologies. The analysis here looks across the issue areas of trade, defense, and agriculture to examine ways in which biotechnologies are being used, the impacts of these uses on relations among forms of power, and the consequences of their use for international relations. It concludes with a look at what this suggests for analysis of the other entrant in the class of meta-technologies, digital information technologies.

INFORMATIONAL META-TECHNOLOGIES, INTERNATIONAL RELATIONS, AND GENETIC POWER: THE CASE OF BIOTECHNOLOGIES

The features that make information technologies so crucial to the issues, institutions, practices, and outcomes of international relations today are the characteristics that define them as meta-technologies, analytically distinct from the industrial technologies that have dominated earlier periods of human history. Meta-technologies vastly expand the degrees of freedom with which humans can act in the social and material worlds. While industrial technologies use a limited range of inputs, often singular, in production processes of limited and often singular mode that produce a limited range of outputs, often singular, meta-technologies can handle multiple and multiplying types of inputs into production processes that are infinitely variable and thus produce an essentially infinite range of outputs.

This change in human productive capacity is both qualitative and quantitative in nature. It is accompanied by a loosening of historical constraints on decision-making about production processes. Some types of path dependency and structural constraints can now be sidestepped altogether. Because the range of possibilities is so much greater than before, what has been learned in the past about how to make decisions does not always suffice. The underlying premodern and modern assumption that an equilibrium can be achieved—that there is a right answer—is irrevocably gone.

In the world of digital information technologies, these characteristics translate into features such as interoperability, mutability, and a rate of innovation that makes ephemerality a commodity in itself. Biotechnologies, however—those technologies involved in the collection, processing, distribution, and use of biological information (DNA)—are also meta-technologies. The history of the information society, going back to its beginnings in the mid-nineteenth century, involved not only the much-discussed innovations in human communication technologies, but also the development of biotechnologies.

In domains of negotiation as historically distinct as those of trade, defense, and agriculture, the meta-technological qualities of biotechnologies have meant that new types of players as well as interactions among them have become critical to the functioning of the international system. Looking at the impact of biotechnologies on international relations in all of these areas makes it possible to see commonalities across them. These commonalities are important and distinct enough that they can be said to comprise the beginnings of a global information policy regime. Most important among them is a shift in relationships among various forms of power, most notably the rise to dominance of a form of power historically relatively rarely used.

The subjects of biological information and its processing via biotechnologies are relatively new to international relations. Braudel (1977) noted that while historians have long examined the accumulation of mineral and human resources, almost no attention has been given to plants. Even among those who have written on the narrower question of the impact of innovation on global agriculture, few have looked at questions raised by new plant varieties (Kloppenburg 1988). Yet microorganisms were first used as a technology, for leaching metals from low-grade ore and for fermenting breads and drinks, as early as 7000–5,000 B.C.E. (Krimsky

1991). Lewis Branscomb (1993) and Poitras (1997) have begun the work of linking analyses of digital information technologies with those on biotechnologies at the most metaphoric level; the analysis here should advance the discussion by delving more deeply into the subject and by doing so within the context of the theories of power that are central to the practice and understanding of international relations.

As with digital information technologies, the use of biological information technologies in recent decades has had significant impact in international relations. Biotechnology has influenced the contents and directions of international trade; redistributed relative weights among both factors of production and classes of players; been a significant force at the forefront of the commoditization processes that remain important to capital; and forced attention to matters of knowledge production, the role of science, and risk within diplomatic circles. Understanding these impacts, and their bases in meta-technological characteristics, should not only be useful in its own right but also will contribute to understanding the effects of digital meta-technologies. Holding both types of meta-technologies in vision simultaneously is increasingly important because, in the current phase of the ongoing convergence of technologies, those that process digital information and those that process biological information are coming together in fields such as bioinformatics (the use of biological matter to process digital information in computing), artificial intelligence (the effort to mimic human intelligence through the use of biologically-based concepts and techniques such as those of neural nets), and biotics (the autonomous evolution of nonhuman life forms in digital space). In the hard world of law, the effects of this convergence are already being felt in areas such as intellectual property rights that are important to the contemporary international relations agenda.

The chapter opens with a brief introduction to the concept of genetic power, goes on to review the impact of biotechnologies on three domains of international relations (trade, defense, and agriculture), and examines commonalities in the use and effects of the use of biotechnologies across those three realms.

GENETIC POWER

Political scientists have long focused attention on power in three forms:

instrumental: power that shapes human behaviors by manipulating the material world via physical force;

structural: power that shapes human behaviors by manipulating the social and material worlds via rules and institutions; and

symbolic: power that shapes human behaviors by manipulating the material, social, and symbolic worlds via ideas, words, and images.

The notion that "knowledge is power" is, in this typology, incorporated into analysis of instrumental, structural, and symbolic power. The easy equation of symbolic power with information—and the relative ease of conducting research into the content of symbolic mass communications—has meant that this type of political communication has received by far the greatest attention.

In today's highly information-intense society, however, it has become clear that information is not only a distinct form of power in its own right, but has moved to the center of the stage, dominating the uses of all other forms of power and changing how other forms of power come into being and are exercised. The term *genetic* well describes this form of power, as it is addressed at the genesis, the informational origins, of the materials, social structures, and symbols that are the stuff of power in its other forms. In doing so it simultaneously extends power over the noetic universe as well. It can be added to the above typology this way:

genetic: power that shapes human behaviors by manipulating the informational bases of the material, social, symbolic, and noetic worlds.

Genetic power is a particularly important form of power today because it is that which takes the greatest advantage of the distinct characteristic of this stage of the information society, the harmonization of systems—of nationally-based information and communication systems across geopolitical boundaries, of different types of information and communication systems with each other, and of information and communication systems with other types of social systems (Braman 1993, 1995). In a harmonized environment, information flows have the structural effect historically the domain of law; thus the ability to shape those flows and the information they hold, genetic power, is the most important form of power. It is what Lessig (1999) talks about in hypothetical US-based detail in his popular *Code and Other Laws of Cyberspace*, Dezalay and Garth (1998) analyze in its international negotiational expressions, and Lewis Branscomb (1993) refers to when he notes that both digital information technologies and biotechnologies are of strategic concern.

The study of international relations is often confined, as are the negotiations it analyzes, to single issue areas, each with its own history, modes of argument, value hierarchies, and operational definitions. But here, as in the effort to distinguish industries, or genres, or geopolitical borders, it is the characteristic of the postmodern condition that the lines have blurred. In the international arena, as in the domestic, the outlines of emergent information policy regimes thus become most evident by looking across them. Here, the impact of biotechnologies in three domains of international relations—trade, defense, and agriculture—is examined as a way of discerning the commonalities and resonances across them that mark the features of an emergent international information policy regime.

BIOTECHNOLOGY AND TRADE

The theories behind international trade and the stuff of that trade began with biological information at one end of the history, and issues arising out of its treatment dominate the contemporary international trade agenda at the other.

HISTORY

Triggered by developments such as the British Corn Laws, economists Smith and Ricardo launched a theoretical justification—comparative advantage—that still endures for the practice of centrally collecting biological information from around the world and redistributing it according to a design that it was believed maximized capital accumulation for those in the center. These activities were only some among those that led Richards (1993) to describe the imperial nation-states as entities more successful at control of information than they were at control of either land or populations. They were important, though, as the then-new sciences of biology and botany contributed significantly to both ideas and efforts. Not only did biology and botany identify types of information of economic interest, but they also introduced new approaches to experimentation and statistical analysis that were critical to states' growing informational armamentarium. (Strikingly, the decline of empire or a state can have an immediate material impact, as in the case of one of the world's most important genebanks, the Vavilov Institute, and others in the former Soviet Union that have gone into dangerous decline since the Soviet government collapsed [The Economist 1994].)

Biological information played such a key role among resources in the imperial enterprise because it is distributed so unevenly about the globe. Centers of biological diversity—that is, regions rich in a diversity of biological information resources known as Vavilov Centers—are almost exclusively located in regions of the developing world. Starting with the trips of Columbus between hemispheres, flows of biological information so vastly increased in the late fifteenth century it became known as the "Columbian Explosion" (Kloppenburg 1988). As with other types of resources, biological information has been extracted in raw form from societies in dependent relationships with the center, processed and value added in the center, and then returned in value-added form—if at all—to those in the developing world at high cost. In a contemporary example of this process that has played heavily in international trade negotiations of the past decade, biological information from Brazilian rainforests has been used by transnational corporations to manufacture valuable medicines now so protected by intellectual property rights held by those corporations that they are not affordable to the Brazilians from whom the original resources came.

Decisions made in the imperial centers about each colony's niche specialization may have been designed to enhance comparative advantage, but they also wrenched indigenous cultures, agricultural practices, social systems, health care, and ecosystems. Complex multicultures in which humans had coevolved with the other organisms in their natural environments were replaced by monocultures devoted to single crops that were useful for export purposes but often failed to adequately sustain either humans or other organisms in those environments. Decisions made for the purposes of maximizing gains to be made from international trade thus had enormous direct environmental impact.

That this international division of labor did not serve all even within the states of the center was noticed early. British farmers first rioted hundreds of years ago in protest over the forced extraction of food from rural fields to feed those in the cities, leaving farmers themselves without. The enforcement of monoculture, too, was not limited to the production end. The laboring populations of Europe found they had no choice regarding what they ate as they were forced to abandon their traditional grains for potatoes and corn, foods that provide so many more calories per acre than those traditionally grown in Europe that it supported a doubling of the population in the century following their introduction (Kloppenburg and Kleinman 1988). This resonance between disadvantaged classes within the most advanced nations and interests in the developing world continues. It was a striking feature of debates during the Uruguay Round of General Agreement on Trade and Tariffs (GATT) talks (Braman 1990) and characterizes early 2000 struggles over the World Trade Organization. Successful establishment of the division of labor for the processing of biological information required attention to each stage of the processing chain—collection, storage, processing, establishment of property rights, distribution, and use—on several fronts. The various activities of nation-states had the effect of building a global network of biological information and information technologies much like the global telecommunications infrastructure, itself comprised of lines and nodes.

There are two levels of nodes in the biological information network. The first tier was established simultaneously with the building of the first global telecommunications network, the telegraph system run by the British, in the nineteenth and early twentieth centuries. Archives were established in the major cities of Eastern and Western Europe and Russia as gardens, seedbanks, genebanks, or culture collections. A second tier of nodes was established after World War II in centers of biodiversity throughout the developing world via an international network of organizations driven by interests of the most advanced nations, the Consultative Group for International Agricultural Research (CGIAR). Both tiers developed to help address logistical issues, and different types of processing are done in each—the ex situ centers in the developed world provide sites for development of laboratory and experimental techniques, and the in situ centers provide opportunities for the study of biological information in its environmental context. Each has strengths and weaknesses—ex situ collections involve logistics often so difficult that they are best carried out by military units, while in situ centers have been criticized for being staffed by scientists from the developed world who take away information but leave little behind either in the way of information or skills or technology transfer.

The building of each tier of nodes in the global biological information network was accompanied by rhetorical justifications. During the first round, as was done with the free flow of information beginning early in the twentieth century (Blanchard 1986; Braman 1990), this was accomplished by simultaneously describing biological information as "the property of all humankind" and therefore free for the taking, and as a form of property over which ownership rights can then be asserted by those who take. The intellectual property law system that was developed to hold this together was sometimes surprisingly concrete: the US Patent Office itself was one of the most aggressive agents for the collection and redistribution of biological information in the world in the late nineteenth and early twentieth centuries.

THE CONTEMPORARY SITUATION

Many of the factors that have historically affected decision-making about the distribution of biological information—transportation, geopolitical stability, nature of the labor force, available technologies, cultural factors—are path-dependent and sensitive to structural forces. The use of recent meta-technological innovations to handle biological information, with the increase in degrees of freedom they bring, is increasingly disrupting centuries-old patterns by making it possible to substitute one agricultural input for another. In so doing, the most advanced nations are once again destroying the agricultural base of one after another developing country whose global economic niche to begin with had been a part of imperial design.

Several strategies have been used by developing countries to respond to these changes, beginning with the calls for knowledge transfer that were a part of the New World Information Order conversation in United Nations Educational, Social, and Cultural Organization (UNESCO). The reshuffling of the map in international trade, and the rules by which the map is drawn, is taking place today in a number of intertwined arenas that include international trade, intellectual property rights, and shifts in innovation and diffusion strategies. The loss or decline of some of the world's greatest genebanks in the countries of the former Soviet Union and acceleration of the loss of biological information diversity that is the result of environmental destruction lend some urgency to these efforts.

Alarm regarding the future of agriculture in the world trading system was raised as soon as the United States broached the notion of extending international agreements regarding trade in goods to flows of information, or "services" with the first efforts to the Uruguay Round of GATT talks in the 1980s. Aware that the United States and others of the most technologically advanced nation-states were seeking to improve conditions for their newly-important economic sectors, countries in the developing world feared that doing so would only worsen already-poor trading conditions for developing country economies dependent upon agriculture. These concerns remain on the table under the institutional conditions that emerged from those talks, including now the GATT, General Agreement on Trade in Services (GATS), and TRIPS, and the WTO as an organization to implement all of the agreements. While biological information, technologies for its processing, and the products of its processing are all affected by these broad-ranging agreements, in 1999 an additional multilateral agreement was added to the bouquet specifically to protect this type of informational resource, the Multilateral Regime for Genetic Resources.

Intellectual property rights law has been developing in multiple directions in response to demands of quite different types of parties concerned about the preservation and use of biological information. On the one hand, the ability to assert property rights over different types of life forms has been steadily expanding since the early twentieth century. On the other hand, there has been a slowly rising tide of efforts to develop a theoretical, rhetorical, legal, and institutional framework for the protection of traditional knowledge and cultures. Costa Rica led the way in experimentation with contracts between governments and transnational corporations regarding the terms under which biological information could be removed, and ensuring at least some return for the nation-state on the profits generated from that information by pharmaceutical and other companies. Governments are also negotiating with indigenous peoples within their own borders for access to both the biological information on lands over which those peoples have control and to the knowledge those peoples have about how to access and use that information, as the United States has done with the Blackfoot (Ruppert 1994). While it is certainly a positive sign that such arrangements can be achieved, to date the funds involved have generally been shockingly low.

The boundaries of intellectual property rights law are being expanded to permit those rights to be held by communities as well as individuals, and a variety of different approaches is being taken to try to ensure that traditional, as well as "new," forms of knowledge can be protected. Just recently, the Multilateral Agreement on Plant Genetic Resources included a provision to protect farmers' rights. Acknowledging the critically important role that farmers have played in working with the DNA of plants over thousands of years to maximize the utility of that biological information for humans, this provision insists upon both property and decision-making rights of farmers on the basis of these contributions.

Another important policy innovation, first suggested and operationalized by the research center nodes (CGIAR) in the global biological information network, was defining a public domain for biological information. CGIAR has established rules for identifying which kinds of biological information should be in the public domain—the basic food grains, those plants upon which the survival of humankind depends, in the forms of farmers' varieties and landraces, obsolete varieties, advanced lines, genetic stocks, and wild species—and has puts its own information of this kind into the hands of the UN's Food and Agriculture Organization (FAO) for safekeeping as part of its program on Plant Genetic Resources for Food and Agriculture. In 1998, CGIAR called for a moratorium on granting intellectual property rights on this "designated plant germ plasm," claiming it should be held "in trust for the benefit of the international community, in particular the developing countries."

Nation-states themselves are increasingly claiming to hold intellectual property rights over biological information, thus presenting a completely different policy profile from that presented for human communications and information. In some cases this is done by fiat (as the South African government has done with an AIDS vaccine), and in some cases by law (as when the U.S. Patent and Trademark Office gives a government agency, the National Institute of Health, patents for all ex vivo techniques used in gene therapy). Governments also negotiate terms for the collection and processing of the biological information of their human populations, as in the case of Iceland, which has agreed to permit scientists to analyze the genetic makeup of the entire population. (The Human Genome Diversity Project is collecting the biological information of a number of peoples defined as relatively isolated, a practice called "harvesting" their diversity [McNally and Wheale 1996].)

Other developments have been strategically conceptual. The concept of food security has been linked to notions of national security, a move that has been successful in raising the salience of policies dealing with biotechnologies on the global agenda. While in the 1980s agriculture was pitted against the information economy, today's agricultural organizations and corporations emphasize the fundamental dependence of the entire global economy upon biological information. As it says on the website for the FAO's Commission on Genetic Resources (www.iisd.ca/linkages/biodiv/comm7.html), "Plant Genetic Resources for Food and Agriculture (PGRFA) are the biological basis of world food security and, directly or indirectly, support the livelihoods of every person on Earth." Innovation strategies have also changed. They have come in response not only to political criticism, but also to repeated experiences of devastation wrought by the vulnerabilities of agricultural monocultures—one of the first actions of the United States upon the fall of the Soviet Union was to rush to Russia for genetic stocks needed to recuperate from a devastating blight on the monocultural U.S. corn crop. The new approach to development of crops for sustainable agriculture of a kind that can best address the world's food needs is called "base-broadening." While during the modern period biological information was sorted and processed in geographically centralized sites for redistribution around the globe according to what was determined to maximize comparative advantage, the postmodern approach to the collection, processing, sorting, and redistribution of biological information is to centralize it not geographically, but within the seed itself. The research effort now is to find ways of incorporating all of the potential advantages of a genestock available for a species into seeds to be distributed locally for differential adaptation to local contexts.

SUMMARY

The collection and redistribution of biological information was central to the imperial enterprise, and thus to the structuration processes shaping the basic geography of international trade. In this, biotechnology has relied upon and contributed to the exercise of power in its instrumental, structural, and symbolic forms by nation-states. The increase in the meta-technological characteristics generated by biotechnological innovation in recent years, however, has destabilized the global structure as actors of a wide variety of types—many historically without any real power in the international or, often, local arenas—learn how to use the genetic power now newly available. The results of this shift for international relations are three fold: there is an increase in turbulence and uncertainty in the international trade system, attention of the international trade system has turned from product to process, and new modes of action and types of agents have been enabled by recent biotechnological innovation. Biological and digital information technologies are now converging in the area of international trade not only in what is traded but in the revamping of the intellectual property rights system in such a way that changes directed at problems raised by either will be applied to both.

Nation-states are players in this process in four different ways. They can contract for ownership of or control over domestic biological information resources with those local entities that have historically had control. They can simply assert national ownership of and control over unprocessed domestic biological information resources. They can assert control over biotechnologies and their products that are invented by their citizens. Finally, the biological information of the citizenry itself can be treated as a resource.

BIOTECHNOLOGY AND DEFENSE

The use of biological information as a destructive weapon is ancient, and the eruption of conflict over access to biological information (at least in the form of reproducible food) even older. Over the past couple of centuries, biotechnologies have also been important to the logistics and economics of war. Today, tensions over access to biological information and the value to be added from its processing are so keen that there is talk of "seed wars" and "gene wars." Advances in the ability to process such information have been important in improving the capacity to conduct war, providing not only weapons but also logistical support and tactical and strategic diversity. The mega-industry that has developed as a result of mergers among corporations in the previously distinct chemical, pharmaceutical, agricultural, food, and brewing industries is today part of a network of defense producers and suppliers that includes those in the nuclear and ballistic missile industries as well (Frankel 1991).

Biological and digital information technologies are now directly linked. The US defense establishment is now encouraging research that will permit co-design of soldiers and the weapons they use, the former involving biotechnologies and the latter digital information technologies intended to complement each other, and the Japanese government has announced that it is holding off on using software commissioned to support government functions and services because they discovered some of the programming had been handled by the same terrorist group that had launched a biological weapon on a subway. The movement from metaphor to reality could not be more clear than here—the Japanese government fears a digital explosive contagion of the same kind that had been experienced using biological information.

History

Biotechnologies have been important to war economically by providing direct economic support, serving as an affordable and accessible source of resources, and making import substitution possible (Bud 1993; Kloppenburg 1988). In turn, military spending on R&D in the area of biotechnology is believed to have a trickle-down effect in terms of innovation that is economically important to the domestic economy as a whole during peace as well as war. De Landa (1991) goes so far as to argue that agriculture itself first evolved out of defense needs, when the walled city was built and then needed systematized agriculture to feed its population. A different type of trickle-down may be noted in Japan, a country that had been using biotechnologies to treat sewage as early as 1914 and to produce ethanol by 1935—but which also began to experiment with the use of radioactivity to process DNA for commercial purposes from the mid-1940s on (Krimsky 1991).

The use of biological information in raw or processed form has provided high drama: The Tatars were the first to use disease as a weapon in the fourteenth century by catapulting dead bodies over city walls (van Creveld 1991), a practice continued during peacetime by Europeans who diffused smallpox via blankets throughout the indigenous populations of North America (Wiegele 1991). The invention of pasteurization transformed warfare by vastly extending its logistical range and led to the granting of the first patent ever on something living, given to Pasteur (Macksey 1989; McKibben 1996). By the early nineteenth century, nation-states had begun to support research in this area out of interest in the potential of biotechnologies to the development of new weapons; this was successful—the first biotechnology-derived explosives appeared in the 1840s, about the same time as the telegraph (Pearton 1984).

The first use of biotechnologically-derived bacteria as a weapon came during World War I, when the Germans used anthrax against Allied livestock (van Creveld 1991). Chaim Weizmann raised economic support for Israel long before it was a nation-state by offering to commercialize the biotechnological processes he had developed during World War I for the production of explosives (Bud 1993). Since that time, the Japanese, British, Americans, and others have been accused of, or admitted, to the use of biological weapons (as often discussed in *Arms Control Reporter*).

Food power is another way in which biological information has been important as a tool of war. The concept of the right to food entered political discourse as early as the French Revolution. Food power had what Paarlberg (1985) describes as "historic credibility" from its importance in Britain during the mercantilist era. Attention to the use of food as a weapon revived in the early 1970s when the United States began to sell grain to the U.S.S.R. in order to prevent mass starvation during the oil crisis, though, as recently as Reagan's administration, the withholding of food from a population in order to achieve military ends was still held in moral disdain by most.

THE CONTEMPORARY SITUATION

In 1986 the Department of Defense identified as areas in which biotechnology might be of use in the future development of new highly infections and toxic viruses, modification of the DNA of target populations as a way of altering the biological functioning of individual organisms, altering the immunological characteristics of diseases for which vaccines already have been found, and bringing toxins into use that have long been known but weren't until recently feasible to use because it hadn't been possible to manufacture them in sufficient amounts. By 1999, the DoD had added to this list modification of soldiers themselves in a process of co-design of soldiers and weaponry. The use of biotechnologies as weaponry remains difficult from the peacemaking and peacekeeping perspective because of the difficulty of verifying compliance with arms control agreements when the weapons under discussion are biological and chemical—disarmament within the community of molecular biologists means biological containment (Rabin 1987).

For national security reasons, the export of biotechnologies is restricted; export controls had been specifically applied by the early 1990s to biosensors and biochips (technologies that process biological information for computing purposes) (Wiegele 1991). Technologies that deliver biological weapons are also of concern (Branscomb 1993). These types of restrictions exacerbate tensions raised in the trade arena between nations in the developed and developing worlds.

Technological innovations make a difference in the utility of biology-based weapons because they provide a means of addressing historic problems in their use: the need for sophisticated manufacturing, testing, and storage capabilities for materials that are often fragile and ephemeral; a delivery system; and the ability to protect one's own troops and population (Wiegele 1991). Still, there are problems: biological weapons are unpredictable once released and often undetectable in production and use, but are relatively transportable and needed only in very small quantities. For the same reason, however, they are particularly appealing to small terrorist groups of limited logistical or technological capacity.

Those who oppose the use of genetically-modified organisms (GMOs, or GM) in foods believe that this use of biotechnology by corporations is in essence an act of war against farmers, who have not only lost their means of production but also the cultures within which those means of production were embedded and which they in turn sustained. They also believe it is an act of war against consumers, who are losing the opportunity to maintain their health through diet. Growing awareness of the environmental dimensions of national security—the notion that environmental health is the most fundamental need in terms of survival—has also encouraged increasing attention to the technologies through which biological information is collected, processed, distributed, stored, owned, and used.

SUMMARY

Biological information and the technologies that process them have been important throughout human history both as causes of war and as tools used in the fighting of war. While the most obvious uses in both domains involve the use of instrumental power, the value of biotechnology in increasing state capacity in areas like logistics—how far across space and time military efforts can be extended—has meant that these technologies have also been useful in the exercise of structural power. (Structural power, in turn, has been used to stimulate biotechnological innovation.) The deep linkages of certain biological information with national and cultural identities have provided an additional symbolic dimension to the role of biotechnology in war historically.

Today, the increase in degrees of freedom with which inputs can be transformed into outputs made possible by the meta-technological characteristics of biotechnology has increased the utility of biotechnology for the exercise of genetic power as well. Wars, whether psychological, political, or military, are being fought not over access to information but over the refusal to accept a fundamental transformation of biological information, in the debate over acceptance of genetically-modified foods. Talk of gene wars and seed wars fills newspapers. The military situation is considered far less stable than it was during the Cold War not only because of the diffusion of nuclear capacity but because of the new possibilities of information warfare using both digital and biological information and the technologies that process them. Meanwhile one cutting edge of military R&D looks to co-design of soldiers and weaponry, using both technologies that process digital information and those that process biological information.

BIOTECHNOLOGY AND AGRICULTURE

Geographic flows of biological information for agricultural purposes are as old as human society; indeed, some claim language itself arose to serve the need to share knowledge of how to work with biological information. It is fundamental to the most intimate and necessary of human activities—what we eat, where we live, how we stay healthy. After a long slow curve of innovation that began with the use of yeasts to ferment drinks and leaven bread over 9,000 years ago (and perhaps deliberate plant breeding longer ago than that), the study of biotechnologies as a science arose over the course of the nineteenth and twentieth centuries at the intersection of the scholarly disciplines of biology, chemistry, and botany, and of the industries of chemicals, pharmaceuticals, agriculture, food, and zymotology (brewing). Discoveries of the last few decades have enabled an explosive growth of innovation in this area just as digitization of information did for human communication and computing systems. By 2000, products that resulted from the use of these were so widely distributed that concern over them provoked mass political activity in cities around the world. The fact that they are often hidden in products in imperceptible ways fuels the social fears and speculations that led Ulrich Beck to describe contemporary society as a "risk society" (1992). Meanwhile evidence rises that unintended side effects of the use of biotechnologically manipulated germ plasm are often harmful and potentially disastrous, as in the death of monarch butterflies who eat pollen from genetically manipulated corn.

HISTORY

The Columbian Explosion and establishment of the international division of labor in the production of biological information marked a first stage of the use of such information to dramatically alter the nature of agricultural practice, product, and culture for a large proportion of the world's populations, though earlier migrations of people and thousands of years of long-distance trade had transformed practices, foods, and products on another scale. By World War II, plant breeding was perceived as an explicitly political tool of American foreign and economic policy (Kloppenburg 1988).

A second stage was launched by the Green Revolution of the 1960s, when US-developed genetically specialized seeds were distributed throughout the developing world. While the Green Revolution was motivated at least in part by the humanitarian goal of helping to feed the world's hungry—itself serving the geopolitical goal of ostensibly helping peoples in the developing world resist Communism—it also served to further extend monocultural practices. The result served well the economic interests of the US agricultural industry.

Most profoundly—in a story well told by Kloppenburg in his book, *First the Seed* (1988)—the Green Revolution succeeded in commodifying the biological information of the most fundamental human crops by developing seed that would not reproduce. Without the ability to store seeds from one year's crops to use for the next, societies are forced out of self-sustenance and into the global cash economy, and societies, as cultural geographer Carl Sauer warned would happen early on, are destroyed. While some seeds developed in recent decades through the use of biotechnologies may reproduce, they will do so in ways that do not reproduce the desirable characteristics of the original. Over time, more and more inputs into the agricultural process, including herbicides and pesticides, have been incorporated into plants themselves via genetic modification through a process of "bundling" that is the reverse of the "unbundling" that has characterized treatment in recent decades of digital information technologies and the services they offer. Agriculture itself has become increasingly information-intensive on the machine side as well throughout the twentieth century. The highest levels of telephone penetration in the United States in the early decades of the twentieth century was in the midwestern farm states of Minnesota, Iowa, Wisconsin, and Illinois; farmers unafraid of machinery and desperate for communicative capacity set up their own lines along fences and set up wives as operators in farm kitchens.

THE CONTEMPORARY SITUATION

Both internationally and domestically, while some are resisting genetically-modified foods, many farmers are instead seeking the right to use, reproduce, and sell the seeds to such foods. Battles over whether or not farmers have the right to plant or sell that genetically-modified seed that does survive are so intense that they have already risen to the US Supreme Court (*Asgrow Seeds* 1995). This issue is likely to be revisited legally at both national and international levels.

Recent biotechnological innovations have also contributed to change in the structure of the agricultural industry. The first industry, agriculture long stood alone. But here, too, as elsewhere in the economic world, global oligopolistic enterprises are consuming enterprise after enterprise and now, industry after industry. Most agricultural business is today conducted as part of the activities of a few global entities in the now-merged agriculture-food-chemical-pharmaceutical industry, with its enormous advantages of economies of scale and vertical integration.

It is biotechnology companies rather than the information technology firms of Silicon Valley that represent the leading edge of the evolution of organizational form, one of the most commented-upon characteristics of the information society. Three ways in which this is so can be identified: relationships between small and large firms; relationships among universities, corporations, and the government; and the development of financial instruments. Biotechnology has led in the development of new types of relationships between transnational corporations and small research boutiques, with the former often financially controlling the latter while simultaneously distancing themselves from the research phase of the R&D process in an effort to minimize risk and maximize whatever it is that makes a genuinely innovative environment. Companies in the biotechnology sector have been among the most adventuresome in the crafting of financial instruments to fund their enterprises. Much of this has been in the area of what might be termed *virtual* finance, dealing with speculative futures and the conceptual vagaries of what are being called *real options*.

Biotechnology has also led the way, by a decade or more, in the redefinition of relationships between government, industry, and higher education that is now unfolding across disciplines—today led as much by dollars for implementing the use of new information technologies as it is by the research funding that has historically served as the linchpin of the relationship. As university-based researchers spin off their own biotechnology research firms—or as universities try to hold to themselves the patents on innovations developed by their faculties—there are ripple effects on the sociology of knowledge, for the circulation of scholarly information through collegial venues is reduced. This is an interesting turn in a history that begins with the federal government handling all agricultural research itself, first via the Patent Office and then, via the Department of Agriculture, that entity spawned in public universities with the establishment of the land grant system in the second half of the nineteenth century. Ironically, however, most farmers themselves could not see the value in this type of agricultural research until it became clear that it was the result of such research that opened up the international market with the Green Revolution. Today, government agencies that may have funded university-based biotechnology research at times find themselves struggling competitively with the universities themselves for control of the intellectual property rights over inventions thus generated.

More recently, farmers were among the first individual users of GPS systems, with equipment built into tractors to enable them to more precisely fertilize and water their fields. They were early users as well of electronic means of tracking sometimes volatile markets.

SUMMARY

The effects of recent advances in biotechnologies on agriculture as a domain of international relations have been several. They have changed the rules of the game regarding the international division of labor, pulling out from under many former colonial societies the economic underpinnings upon which they had come to rely after the lengthy imperial experience. In turn, however, new opportunities for endogenously-driven definitions of a society's agricultural orientation may be opening up as a result of increasing biotechnological sophistication and its concomitant access to genetic power.

Interestingly, there are tensions so strong that they are generating open conflict at both extremes of the spectrum of possible responses to the use of genetically-modified seeds. Some consumers in North America and in Europe are so resistant to the use of such seed that they are trying to shut down international trade negotiations, on the one hand. On the other, the desire of many farmers within both developing and developed countries is so extreme that the issue has already, for example, risen to the Supreme Court level in the United States. Acceptance or rejection of the processing of biological information in certain ways has thus become added to the panoply of tools available to nation-states seeking to improve conditions for their own agricultural producers in the global market, or for those trying to protect certain cultural, health, or environmental positions.

Thus in agriculture, too, it is the meta-technological characteristics of biotechnology that are having the most impact on international relations, and again because these meta-technologies enable the use of genetic power.

BIOTECHNOLOGY, INTERNATIONAL RELATIONS,
AND GENETIC POWER

While the transformational characteristics of biotechnologies that qualify them as meta-technologies have always been there, the extraordinary increase in their meta-technological capacity and the speed at which that capacity is growing today are at the center of fundamental changes in the subject and practice of international relations. In every domain—trade, defense, and agriculture—the use of biotechnologies has been key among the forces that are disturbing equilibria, disrupting centuries-old patterns, increasing turbulence, unpredictability and, some would argue, risk. The global division of labor of the last several hundred years is unraveling as monoculture economies become undermined when the use of biotechnologies enables replacement of commodities they have long produced, thus increasing geopolitical instability.

The recognition that some genetic power is available to nation-states and societies in the developing world as a result of their control over what is by far the larger proportion of biological information resources is leading to a shift in the balance of power between the developed and developing worlds, and within the developing world. While the success of new types of contractual arrangements, the expansion of intellectual property rights, establishment of a public domain for biological information and other recent developments should not be over-stated, these trends do mark a significant—and positive—change in direction for the global system. The need to deal with the products of biotechnological processing has also significantly contributed to a shift in attention in international trading arenas from product to process, echoing a similar shift in the competitive target for those seeking to assert intellectual property rights.

The convergence of technologies that process biological information with those that process digital information is marking each domain of international relations. The evolution of intellectual property rights regime that is the lynchpin of today's international trading system simultaneously affects both types of meta-technologies; indeed, lawsuits are already being pressed that use law developed in response to one type of meta-technology to seek redress for harm caused using the other (e.g., *Florida Prepaid Postsecondary Education Expenses Board* 1999). The U.S. defense establishment is now encouraging complementary research in the use of the two types of meta-technologies, and there is fear that those engaged in information warfare may well do so using both biological and digital information as weapons or subjects of attack. In agriculture, the sites of convergence include the increasing information-intensivity of farming equipment and commodities markets, and it is the use of new information technologies that has enabled the rise of the large oligopolistic firms that now dominate the global pharmaceutical-chemical-agricultural-food industry.

The two earlier transformations in the relationships of society to biological information and biotechnologies resulted from its interactions with capital. First the international division of labor was used to generate capital for those in the imperial center, and then the biological information itself became commoditized. The current period of transformation, marked as it is by a noted and sustained lack of actual profitability on the part of biotechnology firms, has to do with capital of another sort. (Money is, of course, only one form of capital, which is in essence the capacity to make things happen.) Criticism of the way in which the concept of comparative advantage was operationalized during the colonial and early postcolonial periods does not necessarily lead to the conclusion that the concept itself must be abandoned. Though in the past implementation of the concept started from the notion that the design of such a system was better done through centralized planning than through self-organization, it is possible to imagine a global economic system in which each society's niche is defined endogenously rather than exogenously, and according to variables that include social, cultural, political, and environmental goals as well as economic.

The issues raised here should be familiar. They are the same issues that arise as a result of the use of digital information technologies: access to information; the tension between ownership of information and the value added by its processing; and conflicts generated among the multiplicity of functions the same information plays in society, from cultural to economic. The simultaneous study of the two types of informational meta-technologies should enhance our ability to understand the meta-technological effects that they share in common and thus expand the conceptual and policy toolkits we have available to deal with each. Elsewhere, for example, analysis of the treatment of facticity in the realm of biological information reveals a distinction between first, second, and third order facticity issues evident in the treatment of biological information that could usefully be applied to the complex of questions dealing with facticity so centrally important to communications law (Braman 1999).

Because the two types of meta-technologies are themselves converging, the need to keep them both in vision at the same time is growing. This approach is valuable for illuminating not only problems currently on the table, but also those that are not yet emergent but surely coming. Only in this way, for example, does it become evident that the treatment of bundling of different types of information processing is headed in opposite directions as applied to biological as opposed to digital information; while policy is encouraging the unbundling of different types of information processing as applied to the latter, it is encouraging bundling as applied to the former. As with the impossibility of simultaneously applying quite different legal systems to converged communication and computing technologies, it will be impossible for both of these policy positions regarding bundling to simultaneously be applied to converged biological information and digital information technologies and their products.

The capacity for genetic power newly available to both state and nonstate players because of advances in the meta-technologies such as biotechnologies has shifted the balance within and framework for global relations. It is hoped that identification of this fundamental change will contribute to the ability to analyze distinctions among types of power, responses to them, and development of capacity for them.

REFERENCES

Arms Control Reporter.

Asgrow Seed Co. v Denny Winterboer. 1995. 33 U.S.P.Q.2D (BNA) 1430, 115 S. Ct. 788, 130 L.Ed. 2d 682, 63 U.S.L.W. 4055.

Beck, Ulrich. 1992. *Risk Society: Toward a New Modernity.* London: Sage Publications.

Blanchard, Margaret. 1986. *Exporting the First Amendment.* New York: Longman.

Braman, Sandra. 1990. Trade and information policy. *Media, Culture and Society.* 12: 361–385.

———. 1993. Harmonization of Systems: The Third Stage of the Information Society. *Journal of Communication.* 43(3): 133–140).

———. 1995. Horizons of the State: Information Policy and Power. *Journal of Communication.* 45(4): 4–24.

———. 1999. "Are Facts Not Flowers?" Genetic Information and Facticity. Presented to the International Communication Association, San Francisco. (May)

———. forthcoming. Update: From the GATT to the WTO: Snapping in the global trade wind. In Justin Lewis and Toby Miller, eds. *Cultural Policy: A reader.* Oxford UK: Blackwell.

Branscomb, Lewis. 1993. *Empowering Technologies: Implementing a US Policy.* Cambridge: MIT Press.

Braudel, Fernand. 1977. *Afterthoughts on Material Civilization and Capitalism.* trans. Patricia M. Ranum. Baltimore: The Johns Hopkins University Press.

Bud, Robert. 1993. *The Uses of Life: A History of Biotechnology.* Cambridge: Cambridge University Press.

CGIAR. 1998. *Report of the Genetic Resources Policy Committee.* Washington, D.C.: CGIAR, document MTM/98/07.

Commission on Genetic Resources for Food and Agriculture. 1999. *Progress Report on the World's Information and Early Warning System on Plant Genetic Resources.* Rome, Italy: Food and Agriculture Organization, United Nations.

De Landa, Manuel. 1991. *War in the Age of Intelligent Machines*. New York: Zone.

Dezalay, Yves, and Bryant G. Garth. 1998. *Dealing in Virtue: International Commercial Arbitration and The Construction of a Transnational Legal Order*. Chicago: University of Chicago Press.

FAO's Commission on Genetic Resources website, http://www.iisd.ca/linkages/biodiv/comm7.html.

Florida Prepaid Post-Secondary Education Expense Board v College Savings Bank & US. 1999. 119 SCt 2199, 144 LEd2d 575.

Frankel, Benjamin, ed. 1991. *Opaque Nuclear Proliferation: Methodological and Policy Implications*. London: Frank Cass and Company Limited.

Frankel, Otto H. 1988. Genetic Resources: Evolutionary and Social Responsibilities. In Jack Kloppenburg, ed. *Seeds and Sovereignty: The Use and Control of Plant Genetic Resources*. Durham, NC: Duke University Press. 19–48.

Kloppenburg, Jack Ralph Jr. 1988. *First the Seed: The Political Economy of Plant Biotechnology*. Cambridge: Cambridge University Press.

Kloppenburg, Jack Jr., and Daniel L. Kleinman. 1988b. Plant Genetic Resources: The Common Bowl. In Jack R. Kloppenburg, Jr., ed., *Seeds and Sovereignty: The Use and Control of Plant Genetic Resources*. Durham, NC: Duke University Press. 1–18.

Krimsky, Sheldon. 1991. *Biotechnics & Society: The Rise of Industrial Genetics*. New York: Praeger.

Lessig, Lawrence. 1999. *Code and Other Laws of Cyberspace*. New York: Basic Books.

Macksey, Kenneth. 1989. *For Want of a Nail: The Impact of War on Logistics and Communications*. London: Brassey's (UK) Ltd.

McKibben, Bill. 1996. Some Versions of Pastoral. *The New York Review of Books*. New York: McNally & Wheale. July 11. 42–45.

McNally, Ruth, and Peter Wheale. 1996. Biopatenting and Biodiversity: Comparative Advantages in the New Global Order. *The Ecologist*. (Sept.–Oct.) 26(5): 222–229.

The Economist. 1994. Needed: Seed money. Sept. 10 332(7880):100.

Paarlberg, Robert L. 1985. *Food Trade and Foreign Policy: India, the Soviet Union, and the United States*. Ithaca: Cornell University Press.

Pearton, Maurice. 1984. *Diplomacy, War and Technology since 1830*. Lawrence: University Press of Kansas.

Poitras, Manuel. 1997. "Biotechnologies, Commodification and Restructuring." Presented to the International Studies Association, Toronto. March.

Rabin, Robert. 1987. Sustaining American Leadership in Biotechnology. In Indra K. Vasil, ed., *Biotechnology: Perspectives, Policies, and Issues*. Gainesville: University of Florida Press. 229–234.

Richards, Thomas. 1993. *The Imperial Archive: Knowledge and the Fantasy of Empire*. New York: Verso.

Ruppert, David. 1994. Buying Secrets: Federal Government Procurement of Intellectual Cultural Property. In Tom Greaves, ed., *Intellectual Property Rights for Indigenous Peoples: A Sourcebook*. Oklahoma City, OK: Society for Applied Anthropology. 111–128.

Sedjo, Roger A., and R. David Simpson. 1995. Property Rights Contracting and the Commercialization of Biodiversity. In Terry L. Anderson and Peter J. Hill, eds., *Wildlife in the Marketplace*, Lanham, MD: Rowman & Littlefield Publishers Inc. 167–178.

Strobel, Gabrielle. 1993. Seeds in Need: The Vavilov Institute. *Science News*. (Dec. 18 & 25) 144(25–26): 416–417.

van Creveld, Martin. 1991. *Technology and War: From 2000 BC to the Present*. rev. ed. New York: Free Press.

Wiegele, Thomas C. 1991. *Biotechnology and International Relations: The Political Dimensions*. Gainesville: University of Florida Press.

PART II

THE CHANGING SCOPE OF POWER AND GOVERNANCE

CHAPTER FIVE

CIRCUITS OF POWER:
SECURITY IN THE INTERNET ENVIRONMENT

RONALD J. DEIBERT

In 1995, the United States Central Intelligence Agency and Department of Defense issued a joint press release noting that "The security of information systems and networks is *the major security challenge* of this decade and possibly the next century." Given the pantheon of both old and new security threats—from nuclear weapons to environmental degradation—such a pronouncement was of no minor significance. Indeed, in a very short time the Internet has acquired a rather ominous association, one that invokes images of anonymous hackers and crackers, nebulous transnational criminals and money launderers, *cyber-terrorists*, pornographers and pedophiles. At the root of this more ominous association is the belief—articulated in an increasingly large volume of popular and academic literatures—that as societies become more dependent on networked information infrastructures, they also become more vulnerable to potential electronic catastrophe, either through accident or malicious intent. These new problems of security in the context of the Internet are the focus of this paper.

As many have commented, security is a loaded term that activates a powerful set of interconnected symbols and ideas. To be thrust into the realm of security, an issue takes on the imprimatur of utmost importance; the division between the high politics of military security affairs and the low politics of economics reflects this importance. More specifically, the notion evokes a specific set of responses characterized by what Paul Chilton calls "metaphors of containment"—that is, state surveillance of, and territorial defense from, external or outside forces (1995). A residue of the Westphalian war system—where states have been the primary aggregations of political power with territorial encroachment from other states in the system constituting the primary threat—security has been traditionally conjoined with policies of fortification, balancing, and a hardening of the outer shell of the state (Herz 1957). It is because of these associations, and recent policy initiatives by governments in China, Singapore, Germany, and elsewhere, that many foresee a coming government "clampdown" on the Internet.

Yet a quick glance at some of the ways security is being used in conjunction with the Internet reveals a more complex picture. Certainly the steps taken by the Chinese government to build a great "firewall" fall in step with the expectations outlined above, as do attempts by sectors of the U.S. government to limit the spread of enhanced encryption technologies. But out of step with these expectations are ideas concerning networked communications and computer security in areas such as E-commerce or corporate communications. Rather than building walls and clamping down on the Internet, here the emphasis is on devising policies and protocols to further accelerate transnational communication flows. In this sense, security is employed with reference to insuring the validity of purchase transactions, detecting network viruses, and preventing system crashes—measures designed to free up, rather than clamp down on the global information infrastructure.

What, then, does security in the context of the Internet mean for the development of global communications? Is the Internet a security threat? If so, to whom or what is it a threat and in what ways? How will the resolution of the Internet- security problematic affect world order?

A first cut at answering these questions is suggested by several perspectives falling within the rubric of so-called critical studies of security (Krause and Williams 1996; Lipschutz 1995; Huysmans 1998; Williams 1998). Although diverse, together these studies provide two basic analytical points that make them especially attractive to this study. First, they emphasize the historicity of notions of security—that is, that security is not a notion that is fixed and transparent, but something produced in history and changes over time (Krause and Williams 1996, 49).[1] Second, they underscore the constitutive nature of collective images of security (Cox 1986, 218-19).[2] Ideas and theories of what constitute a security threat, in other words, promote and reproduce a particular type of world order by privileging a particular set of policy responses, and an object or referent that is to be secured. Assessments of whether some issue or actor is a security threat, in other words, always presuppose an object that requires securing and a type of political order that is valued. Although the latter has traditionally centered on the nation-state, it need not necessarily be so, and can conceivably encompass other actors or objects in the future.

While these critical perspectives provide a useful framework to assess the various collective images of security in the Internet environment, they are ultimately incomplete. Although they tell us much about the normative content of competing collective images, they tell us nothing about which of them will likely predominate over time. To complete this analysis, we must turn to the material context in which such collective images circulate, compete, and are facilitated and constrained. One perspective that can help illuminate this context is an historical materialist approach to communications called medium theory, developed by theorists such as Harold Innis and Marshall McLuhan (Innis 1952; McLuhan 1964). At the heart of the medium theory is the argument that modes of communication are not mere empty vessels or transparent channels, but significant causal factors that have an effect on what is communicated and how. Changes in modes of communication have important implications for society and politics depending on the nature or characteristics of the technology concerned.[3]

Although this basic proposition has been articulated in different ways by theorists, I have argued that the most useful way to conceptualize it in a nondeterminist way is by drawing out an analogy used by some medium theorists: media as environments (Deibert 1997). Communications environments—defined as the material properties of communication technologies and the political and economic context in which such technologies are embedded—facilitate and constrain social forces, collective images, and ideas much the same as natural environments facilitate and constrain the reproduction of species. As part of the structural-material landscape in which human beings interact, communications environments do not generate social forces, collective images, and ideas *de novo*; they do not impose thought or behavior. Rather, they place obstacles and constraints in the face of some, while providing intensity and dynamism to others. An examination of a changing communications environment can thus help illuminate which collective images will predominate over time and, in doing so, help to trace the changing contours of world order.

The paper will proceed in the following way: Drawing from a framework derived from a critical security studies approach, I will first examine four collective images of security in the Internet environment: national security; state security; private security; and network security. In examining each of the collective images, I will highlight the differing threat perceptions, objects or referents of security, policy responses, and type of world order promoted by each. As will be shown below, the Internet security problematic is not a unified field but a complex intersection of interests and values some of which overlap and some of which collide. To assess which of these competing collective images will predominate over time, in the final part of the paper I outline several properties of the communications environment that constrain national, state, and private while supporting the network security collective images. In the conclusion to the paper, I sketch out what some of the implications for world order might be of these developments.

COLLECTIVE IMAGES OF SECURITY IN THE INTERNET ENVIRONMENT

The Internet phenomenon is well known, even if its social and political implications are not fully understood. Several histories have been written about its origins in the United States military-industrial complex, and the unique architectural principles out of which it has evolved (Hart, Reed, and Bar 1992; Campbell-Kelly and Aspry 1996; Hafner and Lyon 1996). Its rapid growth and increasing penetration into society—facilitated by cheaper and easier-to-use technologies—is now well established with more in store for the future. Such growth and penetration have capped a century or more of radical and fundamental changes in communication technologies, which have re-shaped nearly every aspect of society, economics, and politics on a global scale (Castells 1996).[4] At the end of the twentieth century, we live in a hypermedia environment of planetary digital electronic telecommunications.

Amidst this new environment, bundles of interests and values are coalescing and competing with each other, redefining the boundaries of power and authority on a global scale. Circulating through these bundles of interests and values are several collective images of how security should be conceptualized in the Internet environment. A vast literature has been spawned that extends across several economic, political and military spheres. Overlapping in some areas, while colliding in others, the ideas that inform the Internet security problematic present rival perspectives on the nature and legitimacy of prevailing power relations, and the meanings of justice, public good, and order (Cox 1986, 218–219). This swarm of collective images is the site out of which the contours of future Internet development and, to a significant extent, world order will be shaped.

To help disentangle these competing collective images, an analytical template or framework derived from critical approaches to security is particularly helpful. This framework includes the following questions:

1. In what ways is the Internet seen as presenting a security threat?
2. Who or what is presumed to be the object of security in this regard?
3. What specific policy measures are deemed necessary in response to that threat?
4. What type of world order is promoted and (re)produced by numbers 1, 2, and 3, above?

Four collective images will be assessed using this framework: national security; state security; private security; and network security. It should be emphasized that these collective images are ideal types and not rigid divisions actually existing in practice. States hold positions and elites makes statements that fuse together elements of all four. Additionally, there are potential compatibilities between each of them. As ideal types, however, they help focus analysis on differences, tensions, and contradictions between dominant collective images of security circulating today in the Internet environment.

NATIONAL SECURITY

The historical relationship between the formation of national identities and communication technologies is well known. As theorists ranging from Harold Innis and Marshall McLuhan to Benedict Anderson have observed, the development of mass printing technologies in western Europe was critical in freezing linguistic drift and cementing collective identities around shared vernacular languages (Anderson 1983). Further developments in mass media, such as radio and television, amplified collective cohesion through centralized broadcasting. The integration of mass broadcasting with state interventionist policies over content resulted in a hyper-modern fusion of nation and state. It is for these reasons that throughout the twentieth century such a high premium has been placed, both theoretically and in practice, on state control over mass media as a pillar of political development. Among totalitarian regimes such controls are absolutely enforced and rigidly applied. But even among the most benignly liberal-democratic states, public broadcasting has been widely perceived as a state responsibility to cultivate and preserve a shared national identity.

The emergence of the Internet presents a challenge to this historical formation. Although nations are widely perceived by those who identify with them as deeply entrenched, there is a realization that their vitality is nonetheless contingent on a variety of political protections in the communications field. Language laws in Quebec, television and radio content regulations in Canada, film and video regulations in France and Iran, the outright banning of television in Afghanistan by the fundamentalist Taliban, all attest to the perception that national identity and communication technologies are closely intertwined.[5] As new modes of communication have emerged that are based on principles other than the mass broadcasting paradigm, these types of protections and regulations have increased. Protecting culture and identity has become a critical concern for many states as globalization has intensified.

To date, the major challenge confronted by these regulations has been the proliferation of television and radio channels and the accompanying globalization of the sources of content. While it is debatable whether or not these regulations have been effective, it is almost certain that they will not be in the networked world of the Internet. To understand why this is so, consider content regulations. The Canadian broadcasting act requires private television licensees to achieve a yearly Canadian content level of at least 60 percent overall, measured over the broadcast day, and 50 percent between 6 P.M. and midnight. (As the national broadcaster, the CBC must ensure that at least 60 percent of its program schedule consists of Canadian productions.)[6] Such a system of regulations can be enforced because the channels through which broadcasters operate are a scarce resource requiring government allocation. If a broadcaster defied the content regulations, the broadcasting license could be revoked and the broadcaster would be unable to reach the audience.

In the Internet environment, however, there is no scarce resource equivalent to the broadcasting spectrum requiring allocation of channels. Channels are potentially limitless. Moreover, the audience comes to the broadcaster rather than the other way around. And broadcasters (meaning website producers) are located around the world, rather than within the jurisdiction of Canada. To extend television content regulations into the Internet environment would thus require a transformation of the Canadian state into a planetary totalitarian regime—a remote possibility, however desirable the outcome might be for some.

In the meantime, however, the intent of the regulations is gradually undermined since anyone connected to the Internet can watch whatever television broadcasts are made available over the World Wide Web. Today, the technology is relatively primitive, though vastly superior to what was available just a year or two ago. Dozens of sites offer choppy "realvideo" broadcasts that range from BBC and CNN news to pornography. It is not unrealistic to assume that current trends will continue to the point where thousands of quality-produced television programs from around the world are made available over the World Wide Web. Radio broadcasts are following a parallel trajectory. Should the integration of media continue to the point that the Web subsumes television and radio entirely, the point of continuing broadcasting regulations would seem negligible. The public broadcaster would be but a whisper in an arena of screams.

From the perspective of this collective image, then, the primary threat that the Internet poses is its potential undermining of collective national identities. The primary object of security is presumed to be the nation—the imagined community of people who share a distinct language or ethnicity. Of the four collective images under study here, this collective image is the one with the least visible support. Several countries (or ministries and departments within these countries), such as Canada, France, Iran, Iraq, Germany, Vietnam, China, Syria, and Myanmar have made official pronouncements that showed a sense of concern about threats to cultural identity in the Internet environment.[7] With some, such as Canada and France for example, the sense of concern seems clearly centered on national and cultural identity as traditionally understood. In others, however, notably China, Vietnam, Iraq, Iran, Syria, and Myanmar, the concern with national and cultural identity is difficult to disentangle from a concern with state or regime security, a collective image that will be dealt with in the next section and one that should be kept distinct. Despite the ambiguity of these cases, it is clear that there is a constituency across several countries that views the Internet as a potential threat to cultural security.

The specific policy responses that are forming around this collective image vary widely. Among liberal-democratic states, such as Canada and France, there is a principled reluctance to censor or block out communications with the rest of the world.[8] An important exception is the willingness to censor communications that violate norms of decency, a measure that has been attempted over the Internet with uneven success by the United States, Germany, and others.[9] Apart from censoring indecent communications, the primary policy response appears to be active state support to ensure a national voice has a presence on the Internet. For example, the Canadian Heritage Ministry states as its goal to "increase the creation, production and distribution of high quality Canadian content in both official languages to sustain a strong Canadian presence in conventional and new media."[10] This has and supposedly will entail capital investment in Canadian media industries and the extension of the Internet into more Canadian communities. Numerous other similar examples could be cited as well.[11] The intent of these policies is to provide the nation with financial and technological life-support systems so that it will survive amidst the harsh climate of the Internet environment.

A second policy response among liberal-democratic regimes has been to form "cultural alliances" in order to build up widespread support for cultural protections in trade regime negotiations. For example, in July 1998, cultural ministers from several countries (with the notable exception of the United States) met to discuss strategies to form an international alliance of cultural ministries.[12] Here, the efforts were directed in a more conventional way towards building regulatory fences to control communication flows. The cultural alliances are the novel aspect, though one whose prospects appear dim in the face of more powerful alliances oriented in precisely the opposite direction. Opening the market for trade in so-called cultural products has been the focus of several recent trade negotiations and a major concern of United States trade policy.

Among more authoritarian and conservative regimes, on the other hand, the policy responses veer much more towards the censoring end of the spectrum with, in some cases, complete isolation and containment of the population from exposure to the Internet. Iraq, for example, has banned access to the Internet, calling it a tool of American imperialism.[13] In Myanmar, not only is the Internet outlawed but mere possession of a computer laptop is a criminal offence punishable with a 15 year sentence.[14] Other states have taken a similar route, believing that the best way to protect cultural identity from the Internet environment is to isolate the cultural group altogether from it. To repeat, it is difficult to determine whether such a strategy is more a mechanism for state or regime survival or genuine concern with national and cultural identity. The policy responses are, nonetheless, identical in each case.

In sum, the national security collective image portrays the Internet as a potential security threat to collective identities, with the nation or culture perceived to be the primary object of security. While this collective security image certainly does not dominate the landscape on Internet politics, it has colored the perspectives of several government ministries and countries around the world. The policy options pursued as a function of this collective image have ranged from complete isolation and containment to active state intervention and promotion of national expression on the Internet. The world order promoted by this collective image is a relatively insular system of nation-states.

STATE SECURITY

While the national security collective image may not dominate the world political landscape, one that is gaining significantly more exposure is characterized by a traditional concern with threats to the power and authority of the state apparatus. Particularly in the United States, though having echoes that reach across the world, concerns have been raised about the potential use of the Internet for strategic military purposes. These concerns are embedded in a highly elaborate debate within military intelligence circles—again, based primarily in the United States—about the changing nature of warfare, although the latter issue has far greater scope (Arquilla and Ronfeldt 1997).[15] A second major concern is the loss of state power and authority because of the unique properties of the Internet, particularly the widespread use of encryption technologies. While this collective image thus has several interrelated dimensions, each perceives the object of security to be the state, defined broadly to include the government and the total territorial space, infrastructure, resources, and people under its control.

The first dimension of this collective image sees the Internet as a potentially new medium of warfare in which states are actively planning to operate. Several studies, primarily within the United States security community, have suggested states are actively engaged in military preparations for Internet warfare. In a recent report to the U.S. Senate the director of the CIA, George Tenet, said that China, Russia, and other states have undertaken extraordinary steps to develop an Internet warfare capability.[16] It is difficult to determine the veracity of these reports, however, since threat construction/distortion is a common practice within U.S. military intelligence circles. What is clear is that the United States itself is actively engaged in such preparations, having gone to great lengths to ensure they receive widespread media exposure (Der Derian 1996). Given the extent of financial, commercial, and other interdependencies between states, however, the prospects of two large states actually assaulting each other in full-blown "electronic warfare" seem remote. Scenarios involving stock exchanges being targeted by states with sophisticated electronic tools of warfare are mitigated by the "blowback" that would be unleashed on the initiating state itself, as the ripple effects of recent financial crises in Asia demonstrate. More realistic, perhaps, would be sporadic low-level electronic disruptions undertaken by so-called rogue states, terrorists, and other nonstate actors.

Indeed, numerous and increasing incidents of the latter sort have contributed the most fuel to the rise of this collective image. The most sensational (but least severe) of them have involved the defacing of web pages, including those of NASA, the CIA, and various government and corporate sites around the world.[17] More consequential and disruptive have been the attacks on electronic infrastructures, the spread of viruses, the delivery of malicious coding, and the theft or destruction of data by underground computing groups known as hackers and crackers. Again, precise estimates are difficult to come by because of the nature of the issue area. Both corporations and state agencies are generally reluctant to report incidences of computer intrusions because of the potential loss of confidence in their capabilities. The episodes that have been reported suggest a growing trend, with increasing recognition among government officials of their potential severity.[18]

In February 1998, for example, the U.S. Defense Department reported that it had been the object of a concerted hacker offensive, which turned out to be three teenagers: two Americans and one Israeli. The latter, going under the codename *Analyzer*, claimed that he had access to 400 Department of Defense computers, though officials maintain no sensitive information was compromised or destroyed.[19] In 1997, a Swedish hacker jammed the 911 emergency phone system in west-central Florida.[20] Numerous other episodes could be cited as well.[21] The General Accounting Office reported that the U.S. Department of Defense experienced as many as 250,000 hacker attacks in 1995 alone.[22] The perception is that while such instances have been mostly undertaken by thrill-seeking computer experts still in their teenage years, there is a real possibility that the attacks could become better organized and funded, and directed towards a clear political agenda. Moreover, the scope and scale of the attacks could increase, with major infrastructures—such as stock exchanges, telecommunications systems, air traffic control networks, and other vital conduits—being targeted and crashed. An illustrative episode along these lines was the theft and destruction of Indian nuclear-related information from the Bhabha Atomic Research Centre by antiwar computer hackers in July 1998.[23]

Apart from direct threats to physical infrastructures, a second related dimension of this collective image is the possible loss of state power and authority. Contributing to this perception is the spread of various anonymizing technologies—and in particular publicly distributed encryption software—that undermine the law enforcement and intelligence capabilities of states. The encryption issue is complex, deeply contested, and involves high stakes for several major societal interests. Traditionally, states have monopolized and tightly controlled sophisticated encryption technologies for law enforcement and intelligence purposes. Their relatively greater pools of capital and computing expertise ensured the maintenance of technological superiority over individuals and other private actors.

Gradually, however, developments in computing technologies have led to the widespread availability of increasingly sophisticated encryption systems, many of which are shared freely over the Internet. Fueling this development has been a demand among corporations and businesses driven, in part, by the need to ensure the privacy of their communications vis-à-vis each other, and in part, by a desire to unleash commerce over the Internet—a topic that will be taken up in more detail below.[24] Today, encryption systems are widely available that are practically impossible to break even for states with access to the most powerful supercomputers.[25] For law enforcement and intelligence officials, such developments pose a fundamental challenge to traditional levers of power, particularly various forms of surveillance, such as signals intelligence (SIGNIT). More broadly, they also present problems for the enforcement of state regulations that, in turn, could facilitate organized crime and fraud.[26]

These concerns in the encryption arena are simply one element of a broader threat the Internet poses to state power. For example, the control of information flows in and out of states for ideological reasons, such as that undertaken by China, Singapore, Iran, and others, becomes increasingly difficult. Once connected to the Internet, it is almost impossible to prevent, from a central node, access to information that is available over the wider network. According to Froomkin, "[s]hort of cutting off international telephone service or concluding an international agreement with all industrialized countries to discontinue telephone service with foreign countries that harbor remailers, there is little that one can do keep out messages from any other country, or indeed to keep citizens from sending messages wherever they like" (Froomkin, 1997). Not surprisingly dissident groups within and outside these countries have organized on the Internet providing access to politically outlawed material and reporting violations of human rights.[27] Even among liberal-democratic states, similar sentiments have been raised about the potential broad-based loss of state control over the political agenda to non-state actors.[28]

The policy responses associated with this collective image have varied widely. In response to the possibility of attacks on electronic infrastructures, the United States has taken the lead in focusing on studies, organizational adaptations, and countermeasures.[29] One of the most visible of these was the creation in 1996 of the President's Commission on Critical Infrastructure Protection to assess vulnerabilities and threats to critical infrastructures across all government agencies.[30] Additionally, all of the armed forces have engaged in wide-ranging studies and, in some cases, operational changes to meet the challenges of information and electronic warfare.[31] Numerous other nongovernmental and quasi-public organizations have emerged in the information security area, though their existence bridges at least two of the collective images under scrutiny here (state and network security). Although no other state has gone to the lengths of the United States in this area, other major powers, such as Russia, China, Japan, Great Britain, and France, have followed a similar, albeit scaled-down, course.

In response to the loss of state power and authority—the second dimension of this collective image—the policy measures undertaken have varied as well depending on the state concerned. In the United States and among most other liberal-democratic states, concern has focused on controlling the unlimited spread of encryption technologies that do not permit access for law enforcement. The for-

mula that has been adopted with little success to date has been the push for so-
called key-escrow encryption software as the industry standard—a measure that
has been vigorously resisted by businesses and privacy advocates alike. Although
the specifics of various proposals differ, all key-escrow systems allow back door access for states to encrypted documents and data. States have also begun to take
tentative steps to collaborate internationally on encryption policies under the auspices of the Organization for Economic Cooperation and Development, the G8
(Group of Eight industrialized countries), and elsewhere.[32]

Among non-liberal democratic states, the policy responses have varied depending on the state's interest in global commerce. For those who do have an interest, policy responses have been characterized by a precarious balancing act that
reveals the contradictions of promoting connections to global information infrastructures for economic reasons while maintaining political controls over the
flow of information.[33] Singapore, for example, characterizes itself as an intelligent
island and prides itself on having one of the deepest penetrations of information
technologies in society. Yet it also attempts to maintain vigorous controls over access to certain types of information.[34] In the Internet environment, such controls
have taken the form of strong restrictions on Internet service providers (ISPs), and
punishment and fines for those who are caught violating them. China has responded by attempting to minimize the access points, or nodes, to the global Internet—in effect, creating a national intranet or what some have termed the *Great
Firewall*.[35] Chinese authorities have also passed sweeping regulations similar to
those in Singapore against computer hacking, viruses, the leaking of state secrets,
and the spread of "harmful information" over the Internet.[36] In announcing the
regulations, the Assistant Minister for Public Security, Zhu Entao, said that the Internet "has . . . brought about some security problems, including manufacturing
and publicizing harmful information, as well as leaking state secrets" and that the
regulations were necessary to "safeguard national security and social stability."[37]
Whether or not these types of responses will be technologically effective is, of
course, a separate matter. Among non-liberal democratic states not so concerned
with global commerce, on the other hand, the response has been much more
simply formulated. In Myanmar, Iraq, Syria, and elsewhere, access to the Internet
is either strictly forbidden altogether or very tightly controlled.

In sum, from the perspective of this collective image the primary threat of
the Internet is the way that it facilitates new nontraditional forms of warfare and
violence, particularly from non-state actors and terrorists. A related threat is the
potential loss of state control over information flows in and out of the country.
The primary object of security is the territorial state or government. Policy responses have ranged from attempts to create territorial firewalls that funnel Internet communications through official nodes (China) to coercive pressures on
ISPs and citizens to restrict their access and distribution of information (Singapore)

to the promotion of key-escrow encryption technologies (United States). The world order promoted by this collective image is a system of sovereign states.

PRIVATE SECURITY

The concern with safeguarding privacy has been an integral component of modern liberal thought and practice since at least the nineteenth century, though obviously having important intellectual precursors prior to that time. Its basic thrust has been directed towards the protection of the private sphere from what are perceived as the potentially oppressive public forces of state bureaucracies and mass democracy (Held 1987).[38] The concern with privacy is, perhaps, most visible in the United States experience, with its provisions for individual rights and numerous checks and balances against concentrations of power. However, it is a concern that is reflected in all liberal-democratic states around the world and is generally considered a fundamental human right.

Although perceived threats to privacy are nothing new, advocates around the world have argued that new technologies, including the Internet, have raised the stakes considerably (Bennet and Grant 1999; Agre and Rotenberg 1998). Information about individuals, which at one time might have had to be manually gathered, filed, and stored, can now be digitized and shared among massive computer databases. Moreover, as more and more aspects of society and economy are folded into the hypermedia environment, an increasing amount of personal information is folded in as well. As Lyon puts it, "In numerous ways what was once thought of as the exception has become the rule, as highly specialized agencies use increasingly sophisticated means of routinely collecting personal data, making us all targets of monitoring, and possibly objects of suspicion" (1994).

Today, transaction patterns, periodical subscriptions, health and education records, loan and credit card data, and other types of information help create an electronic profile of individuals that is then shared among businesses and government ministries (Lyon 1994; Gandy 1993). Such information can then be combined with aerial and space-based surveillance imagery to create sophisticated topographical maps, called geographic information systems, that provide electronic profiles of entire neighborhoods on such topics as disease, crime, and income levels (Martin 1991). Many companies offer such services directly over the Internet.[39] On the Internet itself, personal information, such as email addresses and surfing histories, can be captured from surfers by computer programs located in websites, which is then used to generate electronic mailing lists for advertising purposes or to alter site advertisements to match consumer profiles.[40] Coupled with the widespread use of more dispersed centers of surveillance, such as security, handheld video, and web cams, the image that emerges is of a dense electronic cage in which individuals are totally enmeshed and their lives completely transparent (Berko 1992). No wonder, then, that Jeremy Bentham's eighteenth century design for an all-seeing prison, called the Panopticon, has struck such a resonant chord with so many adherents to this collective image (Gandy 1993).

The proponents of this collective image include both state and non-state actors alike. Among many liberal-democratic states, privacy commissioners or ministries have been created that have constructed laws and regulations to protect privacy. Probably the most elaborate of these is the European Data Protection Directive, which went into effect in October 1998.[41] The Directive creates a bundle of rights and protections for privacy that include stringent measures against personal information trade with companies or countries outside of Europe that do not abide by the conditions of the privacy regime. Some countries, like China for example, have no privacy regulations whatsoever, while others, like the United States, have only minimal ones.

In addition to official privacy commissioners, several high-profile non-state actors orbit around the privacy issue as advocates. Numerous transnational non-governmental organizations have emerged that share and publicize information and lobby governments and corporations, including Privacy International, the Electronic Frontier Foundation, the Electronic Privacy Information Center, and the Global Internet Liberty Campaign. These groups act as umbrella networks for the numerous other smaller and more specialized interest groups that share a concern with privacy. The latter range from groups committed to human rights to cyber-libertarians and anarchists. Indeed, a strong anti-authority/antistate streak still looms large in Internet culture, particularly among many of those influential in its early development.

Apart from the official privacy regulations alluded to above, the common policy element that unites electronic privacy advocates is for the complete deregulation of encryption technologies—a move that pits them directly against state law enforcement and intelligence agencies. In what appears to be a paradoxical position, electronic privacy advocates lobby hard against any government attempts to regulate encryption even while arguing that "the genie is out of the bottle" and that the Internet, by its very nature, is immune to state regulation (Brin 1998). Nonetheless, the numerous and detailed webpages maintained by these groups have provided a highly visible touchstone in the ongoing encryption battles. The coordination among these groups—as in the very prominent "blue ribbon" campaign for Internet free speech—is impressive and suggests a formidable collection of interest groups. Other technologies that preserve the privacy of Internet surfing, such as anonymous browsers and various software shields, are also developed and advocated by these groups.

In sum, from this collective image the threat posed by the Internet and other information technologies is the potential invasion of privacy by states and corporations. The primary object of security is the individual. The policy responses emerging out of this collective image include strict privacy regulations and rules that protect personal data and place strictures against how such data can be used as well as total deregulation of encryption technologies. The world order promoted by this collective image is a system of liberal states constituted on the basis of strong human rights and individual privacy protections.

NETWORK SECURITY

One of the more novel perspectives of security emerging in the Internet environment arises out of the increasing importance of networked information technologies for all aspects of postindustrial economics, including transnational production and global finance. New information technologies are inextricably bound up with fundamental changes in the nature of economic organization, from the structure of individual firms to the location of production to the movement and character of money and finances (Castells 1996). As this penetration has increased, and as more and more aspects of society become dependent on networked information infrastructures, a new image of security is emerging that focuses on protecting the networks themselves from systems crash, loss, theft, or corruption of data, and the disruption of information flows.

This network security image has two related dimensions. The first centers on protecting the integrity of data and the flow of information internal to specific businesses and corporations. As corporate restructuring has evolved away from hierarchical organizational structures and fixed locations towards adaptable networks and multi-locational flexibility (a development itself fundamentally bound up with new information technologies), ensuring the rapid and reliable flow of information, as well as the integrity of such flows, has become fundamentally important. Although many corporations regularly lease their own private networks, called intranets, to ensure the speed and reliability of their data flows, there is increasing pressure to integrate internal networks to the wider Internet.[42] Securing the flows has thus become a major concern, particularly as the number of network attacks has increased.[43] Firewalls, virus-protection software, logging and real-time alarm systems, and various forms of encryption and smart card authentication systems have been vigorously developed and applied by corporations.[44] With reliable encryption packages, for example, virtual private networks (VPNs) can be developed which deliver data around the globe using public networks instead of expensive leased lines.[45] To give one example, the General Electric Corp. planned to have, by 2000, "all 12 of its business units purchasing its nonproduction and maintenance, repair and operations materials . . . via the Internet, for a total of $5 billion."[46] To help service these needs, a market for network security has exploded. Site Patrol International Services, for example, provides corporations not only with the relevant firewall and encryption systems but real-time 24-hour network monitoring to track incidences, identify potential security breaches, and provide rapid responses as well.[47]

The second dimension of this image centers on securing flows of information between producers and consumers, a concern that is at the heart of ongoing attempts to commercialize the World Wide Web but is bound up with broader changes in the marketplace towards informatization. Almost all banks, for example, have invested in and promoted electronic access for customers.[48] Partly justified for customer convenience and competitive pressures, but no doubt related also to the potential downsizing benefits as well, most every banking transaction can be done either through electronic tellers, over the telephone, or through computer access with software packages supplied by the banks. Each step, however, has necessitated an increasing investment in security protocols that includes not just software and hardware (modem pools, compact discs, leased lines, secure servers, access control mechanisms, etc), but computer security consultants as well.[49] The widespread use of smartcards, stored value-devices, and other digital credit systems for consumer transactions and other services around the world have also entailed attention to network security protocols and mechanisms as well.[50]

This convergence of commercialization pressures and new information technologies in both dimensions has created a vortex of interest on the Internet. The pressures and expectations surrounding the commercialization of the Internet and World Wide Web have been large. Predictions have been made for several years about an enormous market for Internet commerce emerging, ones that until recently had not been reached (Angelides 1997).[51] The main stumbling block has been precisely the lack of security for transactions. Consumers have been generally reluctant to use their credit cards over the Internet thus stifling E-commerce. To improve security and unleash friction-free commerce, massive investments have been made in encryption technologies and electronic payment schemes. Several electronic cash systems have emerged, such as Digicash, First Virtual Holdings, NetCash, and Cybercash.[52] As a consequence, a marketplace on the Internet is quickly arising, particularly in those areas—such as financial services and software—that lend themselves to networked communications.[53]

In each of the dimensions noted above—that is, in intra-corporate communications and corporate-customer communications—ensuring that information networks function efficiently and without corruption is of paramount importance. The network itself is the object or referent of security. The scope of the network is largely non-territorial, though of course the policy deliberations that concern it are centered in several state jurisdictions. The threats to network security include a wide range of activities, including: programming errors that could lead to systems crashes or vulnerabilities; computer fraud and theft; disgruntled insiders and employees, who intentionally sabotage computer systems; loss of supporting physical infrastructure; malicious hackers or crackers; industrial or corporate espionage; and malicious coding and software, such as viruses, Trojan horses, and worms.[54]

In some respects, this collective image overlaps with the state security collective image outlined earlier. As with the latter, the network security image has focused attention on preventing the illegal penetration of computer systems, the malicious use of computer viruses, and the potential disruption of major electronic-dependent infrastructures, such as stock exchanges or air traffic control systems. The colossal public attention generated around the so-called millennium bug or Y2K issue, for example, was an area of direct overlap between the two. Like the state security collective image, this image has also contributed to the creation of numerous organizations devoted to safeguarding computer and information security, and the allocation of large amounts of public expenditures towards such ends. It is for these reasons that the two collective images are often intertwined in analyses of information security.

Important differences stemming from the referent of security in each case, however, warrant keeping the two collective images distinct. First, with the network security collective image the primary concern is with ensuring the integrity of information flows internal to firms and between firms and consumers. As production processes have diffused across territorial boundaries, and as capital markets become increasingly globalized, these issues have taken on a fundamentally non-territorial dimension. Salomon Brothers Inc., in other words, is concerned with safeguarding its transactions regardless of the specific jurisdictions in which those operations are located. States, on the other hand, are fundamentally concerned with ensuring the security of information infrastructures within a particular territorially delimited space, and only then as a larger function of the protection of the state itself. In some cases, networks in other national jurisdictions might even be the target of disruption as part of interstate competition.

Second, the network security image is fundamentally oriented towards reducing the friction and enhancing the velocity of information flows. The following quotation from an industry periodical shows the double concern with security *and* speed:

> In teaming up with NSTL Inc. . . . to evaluate six leading hardware-based VPN (virtual private network) devices, we found that all were up to the security challenge, able to fend off more than 200 types of attack. And in most cases, managing devices remotely was easy. But performance? It proved problematic, especially for links of T1 (1.544 Mbit/s) or higher. In worst-case stress testing, devices dropped anywhere from 50 percent to 85 percent of offered loads—and for applications that rely on lots of

short packets (like corporate intranets), dropped packets can lead to lots
of retransmissions. Say so long to savings on bandwidth.[55]

The state security image, on the other hand, is concerned with restricting, collecting, and blocking information flows, should such flows been seen as a threat
to the state. The velocity of flows is either incidental, or of a subordinate concern.

The differences are most strikingly apparent in the respective positions taken
on encryption policies. As mentioned above, encryption touches at the heart of
the state's surveillance capacity. It is for this reason that most states' intelligence
and law enforcement agencies have attempted to maintain tight controls over the
export of sophisticated encryption technologies. In some states, the domestic use
of cryptography is tightly controlled as well.[56] Corporations, on the other hand,
have come to view encryption as absolutely vital to ensuring network security in
both senses outlined above—that is, to protect the integrity of their intranet flows
as well as to ensure the security of transactions in the emerging electronic marketplace. It is for this reason that the giants of the corporate and computing world
have invested billions of dollars in developing Internet security protocols, including encryption, and have been at the forefront of attempts to block government
restrictions.

In sum, from the perspective of the network security collective image the primary threat of the Internet is the potential for systems crash, loss, theft or corruption of data, and interruption of information flows. The primary object of security
is the network. Policy responses include the development and distribution of highly
sophisticated encryption technologies, systems of secure access, Virtual Private
Networks, Intranets, and digital immune systems. The world order promoted
by this collective image is a system of highly-integrated internationalized states
embedded within a dense network of transnational communication flows.

COLLECTIVE IMAGES IN THE HYPERMEDIA ENVIRONMENT

The four collective images outlined above circulate as alternative paradigms of Internet organization and, by extension, world order. Which of the four predomi-
nates will have significant consequences for the nature of politics and authority
into the twenty-first century. While the analytical framework derived from a critical security studies approach helps illuminate the normative content of these collective images, it does not provide any clues as to which of them will likely
predominate over the others—a limitation shared among critical, postmodern, and
constructivist approaches to world politics generally. To complete the analysis, we
must turn to the material context—the communications environment—in which
these collective images circulate, compete, and are facilitated and constrained.

Obviously, given the considerable support that exists for each of the collective images above, we should not expect one of them to prevail fully over the others in the short-term. The institutional inertia and material interests surrounding all of these collective images ensure that none will wither immediately. Shifts in world order—though abrupt in historical terms—occur gradually, often spanning generations. Three elements of the communications environment, however, suggest that the network security collective image will thrive over time while national, state, and private security collective images will be constrained:

1. *The packet-switching, non-linear architecture of the Internet environment.* One of the major constraints of the national and state security collective images is the very architecture of the Internet communications environment itself. The Internet is not a single entity, but rather a networked connection among millions of dispersed computers. Each of these computers adheres to a common interconnection standard, known as TCP/IP. This standard enables the use of packet-switching, which is how information is transmitted through the Internet. In packet-switching, messages are broken up into discrete units, or packets, that are then routed through the network and reassembled once they reach their destination. With packet-switching technology and the distributed TCP/IP network, the data that comprise a single message take multiple independent routes to reach their destination. Hence the common description of the Internet as a decentralized, anarchic network. The constraint that this architecture presents to the national and state security collective images is that as the network spreads and as communication flows become more dense and swift, the difficulties of filtering out or blocking particular types of information mounts. There are no single choke points or nodes through which all information passes, for example. Nor is there any single route through which particular messages travel. Information is scrambled and distributed across numerous independent trajectories along the network (Dam and Lin 1996).[57] Although it is possible for states to completely detach themselves from the network and prevent citizen access altogether, once they opt to connect, the constraints of the network for censorship and other forms of communication regulation loom large. Certainly coercion, threats, and intimidation are employed—perhaps even successfully. From a technological perspective, however, the architecture of the Internet makes them much more difficult to enforce.

2. *Advanced Encryption Technologies.* Although the packet-switching architecture of the Internet may make it difficult to filter out or censor particular types of information, do not digital computing technologies actually facilitate state surveillance—an integral part of the state security collective image? Certainly the tools of electronic surveillance available to states have grown significantly in recent years, specifically artificial intelligence programs employed in network surveillance systems, such as the American Financial Crimes Enforcement Network, or

FinCEN (Helleiner 1998). In fact, the digital character of information and the ever-increasing computing power integral to the Internet would actually make the job of state surveillance enormously more effective were it not for a second property of the communications environment: the wide dissemination of easily accessible and highly-sophisticated encryption technologies. Once the province of state military and intelligence agencies, the mass popularity of computers and improvements in computing technologies have led to the diffuse development of highly sophisticated public key encryption systems. Today, encryption software with keys in the 1000-bit range are freely distributed over the Internet—a level of sophistication that would be resistant for decades to even the most advanced network of Cray supercomputers at the service of government security agencies. Although states may set regulations that prohibit the use and export of such technologies, the consensus among most is that "the genie is out of the bottle" (Dam and Lin 1996).[58] At best, prohibitions against encryption use and key-escrow schemes are contrivances to buy time in a losing battle. The encryption properties of the communications environment clearly favor the privacy and network security collective images outlined above.

3. *Post-Industrial Global Capitalism.* A further boost to the network security collective image is provided by changes in the global political economy, particularly the transnationalization of production and the globalization of finance. Although the full details of the latter are beyond the scope of this paper, they are well documented elsewhere. What is of relevance, however, is that these changes have generated a large constituency of powerful interest groups who support the network security collective image. Transnational corporations, particularly in the knowledge and financial services sectors such as banking, insurance, telecommunications, and entertainment, not only command enormous sums of wealth, but have a material interest in the development of secure global networks. As their corporate structures move further in the direction of flexible, just-in-time production arrangements dispersed across multiple national locations involving mobile and wireless communications, their dependence on the network rises in importance. This has generated not only a structural pressure on states, but a powerful constituency actively lobbying for the relaxation of encryption regulations and generating a vast market of ever-sophisticated network security products as well.[59] As more states mold their policies according to liberal-capitalist principles and in the direction of so-called knowledge economies (partially as a product of the structural pressures of transnational capital), the constituencies resisting or contradicting the network security collective image wither in importance and influence. Advocates of privacy, though having a largely independent set of concerns, gain in the wake created by this constituency's support of encryption technologies, though not enough on their own to override the latter's hegemony. Moreoever, the overall transparency of the hypermedia environment in areas that

are unaffected by encryption technologies—the widespread use of surveillance cameras, to give one example—places additional constraints on the private security collective image (Brin 1998).

CONCLUSION

In words that apply equally well to this analysis, Robert Cox commented that: "It would, of course, be logically inadmissible, as well as imprudent, to base predictions of future world order upon the foregoing considerations. Their utility is rather in drawing attention to factors which could incline an emerging world order in one direction or another" (1986). The analysis presented above suggests that world order will incline in the direction of a network security collective image. Of course, powerful elements can be identified that favor national, state, and private collective images as well, including the People's Republic of China, the U.S. law enforcement and intelligence community, and privacy advocates. This support suggests that all of the collective images will coexist for some time to come. Given the constraints and supports of the hypermedia environment, however, we should expect the network security collective image to flourish and become hegemonic over time.

What are the implications for states and world order of the rise and flourishing of this collective image? What is most significant is that the policies associated with the network security collective image are oriented in precisely the opposite direction of those traditionally associated with the term security. Rather than erecting walls and hardening territorial shells, the network security collective image suggests a hastening of transnational communication flows. As these flows of information accelerate, and the networked web of communications becomes increasingly dense, the structural pressures on states will increase accordingly. The internationalization of the state, and corresponding paralysis of state autonomy and power, will continue, and even magnify (Cox 1986; Cerny 1995; Deudney 1995). Certainly states have not disappeared, nor should we expect them to. However, they are in the process of being turned inside out—locked-in and interpenetrated by an electronic web of their own spinning.[60]

As dominant security concerns shift their focus to the network, the nature and exercise of power will be transformed as well. Rather than being associated with the control of territory—a space-of-places, in Castell's words—power will increasingly be manifested in the control over a "space-of-flows" (Castells 1989). Regulation, direction, and restriction of the tempo and access to circuits of information, in other words, could become the most significant bases of political power (Luke 1991). Like *crypto-palities* of the new medieval world order, these transnational private security regimes could thus very well represent a new political species arising on the world political landscape. They suggest a dispersal of authority to a much wider domain of non-state and private actors, oriented around new frames of space-time references beyond territoriality.[61]

NOTES

1. As Krause and Williams put it, "To understand security from a broader perspective means to look at the ways in which the objects to be secured, the perceptions of threats to them, and the available means of securing them (both intellectual and material) have shifted over time."

2. "Collective Images" is a term I borrow from Robert Cox. He defines collective images as "differing views as to both the nature and legitimacy of prevailing power relations, the meanings of justice and public good, and so forth. Whereas intersubjective meanings are broadly common throughout a particular historical structure and constitute the common ground of social discourse (including conflict), collective images may be several and opposed. The clash of rival collective images provides evidence of the potential for alternative paths of development . . ."

3. Cf., James Rosenau's views on technology in his contribution to this volume.

4. Cf., see also Mark Zacher's contribution to this volume.

5. Ironically, the same fundamentalist Taliban has its own website, at http://www.taliban.com/index.html. There, one can read why "[p]rohibition of TV, VCR was essential to save the society from destruction."

6. See the Canadian Broadcasting Act, 1991, found online at http://www.crtc.gc.ca/ENG/LEGAL/BROAD_E.HTM

7. See Syria's on Net, and on Guard. *Wired News* (July 10 1998). A spokesman for the Syrian Computer Society said that "Our problem is that we are a traditional society and we have to know if there is something that cannot fit with our society." Ta Ba Hung, Vietnamese Minister of Science, Technology, and Environment, said that "information flow might affect badly the cultural identity of the nation." See Keith B. Richburg, Future Shock: Surfing the Net in 'Nam, *Washington Post.* (November 19, 1995). For Iran, see Neil MacFarquhar, With Mixed Feelings, Iran Tiptoes to the Internet. *New York Times* (October 8, 1996).

8. For a good overview of Canadian communications policy with special reference to the Internet, see Eli Turk and David Johnston, "Competitiveness, Access, and Canadian Content: The Three Pillars of Canadian Internet Policy." (Paper delivered to the Impact of the Internet on Communications Policy conference, December 3–5, 1997, Harvard University, Cambridge, Massachusetts, USA. Found online at: http://www.ksg.harvard.edu/iip/iicompol/Papers/Johnston.html

9. See David Hudson, Germany's Internet Angst. In *Wired News.* (June 11, 1998); Stephen Labaton, Computer Stings Gain Favor as Arrests for Smut Increase. In *New York Times.* (September 16, 1995).

10. See *Strengthening and Celebrating Canada for the New Millennium.* Canadian Heritage Portfolio—Overview of Priorities. (Canadian Heritage 1998). Found online at http://www.pch.gc.ca/mindep/misc/millenium/e-9.html

11. See, for example, France's Chirac Pledges Computer, Literacy Drive. *CNN Online,* (March 10, 1997); French Launch Cyberspace War Against English. In *London Times.* (April 14, 1997); and Victoria Shannon, Online Via the French Connection: It Takes a Global Village. *Washington Post.* (June 16, 1997).

12. See Culture Forum Eyes CNN Rival. *The Toronto Star.* (July 1, 1998). The cultural ministers also discussed plans to create a global news organization to rival CNN.

13. Iraq: Internet Yet Another Tool of American Domination. *CNN Online.* (February 17, 1997).

14. See Joshua Gordon, East Asian Censors Want to Net the Internet. *Christian Science Monitor.* (November 12, 1996).

15. The Future of Warfare. *The Economist.* (March 8, 1997).

16. See A Prelude to InfoWar. *Reuters* (June 24, 1998).

17. NASA Web Site Briefly Closed Due To Hackers. *CNN Online.* (March 7, 1997).

18. Jacques Gansler, the U.S. undersecretary of defense for accquisition and technology, said that teenage crackers pose a "real threat environment" to national security. See Wayne Madsen, Teens a Threat, Pentagon Says. *Wired News.* (June 2, 1998). See Pentagon Reports Cyberattack. *Wired News.* (February 25, 1998).

19. See James Glave, Hacker Raises Stakes in DOD Attacks. In *Wired News* (March 4 1998). The teen, Ehud Tenebaum, was eventually arrested by Israeli National Police for "illegally accessing computers belonging to the Israeli and United States governments, as well as hundreds of other commercial and educational institutions in the United States and elsewhere."

20. Robert Trigaux, Crackers—the bad apples among hackers—find government and business easy prey. In *Toronto Star.* (July 4, 1998).

21. As this paper is being written, yet another hacker incident occurred, this time of the U.S. Coast Guard computer systems by a disgruntled officer. See Laura DiDio, U.S. Coast Guard Beefs Up Security After Hack. In *CNN Online.* (July 22, 1998).

22. Trigaux, Crackers.

23. See James Glave, Crackers: We Stole Nuke Data. *Wired News* (June 3, 1998). http://www.wired.com/news/news/technology/story/12717.html

24. Corporate spying is a major factor in the private development of encryption technologies. For discussion, see Adam L. Penenberg, Corporate Spies. *Forbes Digital Tool.* (April 3, 1998). Online at: http://www.forbes.com/asp/redir.asp?/tool/html/98/apr/0403/feat.htm

25. "In July 1997, it took 78,000 volunteered computers on the Internet 96 days to crack a message encrypted with DES (the Data Encryption Standard), a secret key algorithm that uses a 56-bit key. It is estimated that it would take the same computer resources 67 years to crack a secret key algorithm using a 64-bit key and well over 13 billion times that age of the universe to crack a 128-bit key." Government of Canada. (February 1998).

26. For an overview, see Vic Sussman, Policing Cyberspace. *U.S. News and World Report.* (January 23, 1995); M.B. Gayle, Virtual Chaos. *Washington Times.* (May 8, 1995); Pat Cooper, Organized Crime Hackers Jeopardize Security of U.S. *Defense News.* (October 3, 1994); Michelle Celarier, What a Tangled Web. *Euromoney.* (October 1996).

27. See, for example, Michael Clough, Cyberspace: Why Nations Could Fear the Internet. *Los Angeles Times.* (February 4, 1996); Michael White, Now We're Watching Big Brother. *The Guardian.* (August 3, 1996); Gregory Katz, Zapatistas. *The Dallas Morning News.* (March 12, 1995); and Faiza S. Ambah, Dissidents Tap the 'Net' to Nettle Arab Sheikdom. *Christian Science Monitor.* (August 24, 1995).

28. A particularly illuminating illustration in this respect are the views on anonymous re-mailers contained in Strassman and Marlow (1996). See also Swett (July 1995).

29. See the report Information Security—Computer Attacks on Department of Defense Pose Increasing Risks, *Government Accounting Office,* (May 1996) for an early study and recommendations.

30. See the website for the PCCIP at http://www.pccip.gov/. Its first major report, entitled *Critical Foundations* (1997), provided a glimpse of some of the conceptual difficulties that surround adopting traditional notions of security in the new Internet environment, particularly the blurred distinctions between military, civilian, domestic, and foreign infrastructures. The report stated that "[f]ormulas that carefully divide responsibility between foreign defense and domestic law enforcement no longer apply as clearly as they used to. "With the existing rules, you may have to solve the crime before you can decide who has the authority to investigate it."

31. See Security Team Finds Pentagon Computers Unsecured, *CNN Online* (April 16, 1998) for details on counter-intelligence efforts at the Pentagon and elsewhere.

32. See the report OECD, *Cryptography Policy: The Guideliness and the Issues,* (1997) available online at http://www.oecd.org/dsti/sti/it/secur/prod/GD97-204.htm. See also the report of the OECD Emerging Market Economy Forum: Workshop on Cryptography Policy, (Paris 9-10 December 1997), available on-line at http://www.oecd.org/dsti/sti/it/secur/act/emef.htm; see also the Wassenaar Arrangement on Export Controls for Conventional Arms and Dual-Use Goods and Technologies at http://www.wassenaar.org/.

33. See Leslie Helm, Asia Wary of Being Wired. *Los Angeles Times.* (February 3, 1996).

34. See Darren McDermott, Singapore Unveils Sweeping Measures To Control Words, Images on Internet. *Wall Street Journal.* (March 6, 1996).

35. See Rone Tempest, China Puts Roadblocks on Information Superhighway. *Los Angeles Times.* (September 6, 1996); Steven Mufson, Chinese Protest Finds a Path On the Internet. *The Washington Post*, (September 17, 1996); Angela Li, Complete Control of Internent 'Unlikely.' *South China Morning Post*, (January 11, 1997); and Philip Shenon, 2-Edged Sword: Asian Regimes on the Internet. *New York Times.* (May 29, 1995).

36. "China Issues New Net Controls." *Wired News.* (30 December 1997).

37. *Ibid.*

38. Critiques of mass democracy as a potentially oppressive force that could threaten liberty have a long history reaching back to Plato's *The Republic.*

39. See, for example, Analytical Surveys Incorporated at http://www.anlt.com/. The description of the services is as follows: This innovative and rapidly growing company uses a variety of advanced technologies to convert paper-based maps, aerial photography, tax records and other geo-referenced information into a digital format. Once in a computerized form, the information can be combined in layers to form a geographic information system (GIS), or *intelligent map*. A GIS is a powerful and flexible analytical tool that is easily accessed, analyzed and updated by users in a broad range of decision making processes.

40. See Denise Caruso, As Privacy Grows Scarcer on the Internet. *New York Times.* (June 3, 1996); and Pete Slover, Cyber Crumbs: Do Internet Cookies Leave a Trail That Could Threaten Your Privacy? *Salt Lake Tribune.* (July 31, 1998).

41. See Simon Davies, Europe to U.S.: No Privacy, No Trade. *Wired News.* (May 1998).

42. See David Greenfield, Global Intranet Services: Patchy But Promising. *Data Communications.* (March 21, 1997).

43. More than half the network managers at 205 *Fortune 1,000* companies say they have detected attempted break-ins during the past 12 months. Nearly 60 percent of those who know they have been hacked admit to 10 or more break-ins during the same period. "Security." *Data Communications.* (August 1997) 175.

44. For various examples, see the following: Lee Bruno, Plugging Security Holes. *Data Communications.* (February 1998) 29–32; Lee Bruno, Firewall Protection Without the Pitfalls. *Data Communications.* (March 1997) 31–32; Rodney Thayer, Bulletproof IP. *Data Communications.* (November 21, 1997) 60; Charles Cresson Wood, Logging, Auditing and Filtering for Internet Electronic Commerce. *Computer Fraud and Security.* (August 1997) 16.

45. See Security. *Data Communications.* (August 1997) 176; and Joyce Harvey, The VPN Puzzle. *America's Networks.* (April 1, 1998) 43–47. See also Tina Bird, Building VPNs: the 10-Point Plan. *Data Communications.* (June 1998) 123–132. Bird notes that VPNs can be up to 80 percent cheaper than private leased lines.

46. Cited in United States Department of Commerce report, "The Emerging Digital Economy," (April 1998), found online at: http://www.ecommerce.gov/emerging.htm

47. See the Site Patrol International Services website at http://www.bbnplanet. com/products/security/sitepat.htm

48. Illustrating the financial depth and global scope of such activities, ScotiaBank Inc. of Canada is an international financial institution with $200 billion in assets that services 4 million customers in 50 countries. See Lee Bruno, Banking on Trust, *Data Communications* (May 21, 1998) 43–49 for an overview of its extensive network security provisions.

49. In its transition to electronic access services, ScotiaBank Inc. hired a team of "ethical hackers" who worked from a remote site in Palo Alto, California that staged a multipronged computer attack on the mainframe, operating systems, and Web servers. See Bruno, "Banking on Trust," p. 45. For 48 hours of hacking, the price: $35,000.00. The total cost of ScotiaBank Inc.'s deployment of electronic access services was $2,007,000.00. The market for network security products and services was projected to grow by 70% in 1997. See Charles Cresson Wood, Status of the Internet Electronic Commerce Security Market. *Computer Fraud and Security.* (September 1997) 8. See also J.H.P. Eloff and Suzi van Buuran, Framework for Evaluating Security Protocols in a Banking Environment. *Computer Fraud and Security.* (January 1998) 15–19; Laura DiDio, Private-key Nets Unlock E-Commerce. *Computerworld.* (March 16, 1998) 49–50.

50. See Alan Laird, Smartcards—Is Britain Smart Enough? *Computer Fraud and Security.* (February 1997) 11–15; Ivars Peterson, Power Cracking of Cash Card Codes. *Science News.* (June 20, 1998). As online stock and investment transactions have become more common, security concerns have increased there as well. For one example, see Ellen Messmer, Investment Firm Buys Into Public-Key Encryption. In *Network World.* (May 4, 1998) 57–60. See also Sharon Machlis and Jana Sanchex-klein, Will Smart Cards Replace ATMS? *CNN Online.* (July 30, 1998). http://www.cnn.com/TECH/computing/9807/30/homeatm.idg/.

51. See Gordon Arnaut, The Holy Grail of Internet Commerce. *The Globe and Mail.* (November 14, 1995); and Steve Lohr, The Great Mystery of Internet Profits. *New York Times.* (June 17, 1996). While expectations of a market for consumer transactions have not panned out as fully as some predicted, that for business-business transactions has exploded. The United States Department of Commerce report, entitled "The Emerging Digital Economy," forecasts $300 billion in Internet commerce between businesses by the year 2002 based on current traffic trends. The report is located at http://www. ecommerce.gov/emerging.htm

52. See Alasdair Murray, Digital Money Opens Way to Cashless Global Trading. *The Times.* (January 9, 1996); Neil Gross, E-Commerce: Who Owns the Rights? *Business Week.* (July 29, 1996).

53. See Andrew Allentuck, Financial Services That Delight, Amaze. *The Globe and Mail.* (November 14, 1995); and Vanessa O'Connell and E.S. Browning, Stock Orders on Internet Poised To Soar. *Wall Street Journal.* (June 25, 1996). The Dell Corp was selling as much as $6 million worth of computer equipment and software each day during 1997. See the U.S. Department of Commerce report, "The Emerging Digital Economy," at http://www.ecommerce.gov/emerging.htm

54. For an overview, see CSL-Computer Systems Laboratory Bulletin, (March 1994), found online at http://www.nsi.org/Library/Compsec/compthrt.txt

55. VPNs: Safety First, But What About Speed? *Data Communications.* (July 1998).

56. See the exhaustive survey on state cryptography policies at the Global Internet Liberty Campaign website, http://www.gilc.org/crypto/crypto-survey.html. As the report indicates, Belarus, China, Israel, Pakistan, Russia, and Singapore all maintain tight domestic controls on cryptography use.

57. A U.S. National Research Council report noted: "When an interceptor moves onto the lines that carry bulk traffic, isolating the bits associated with a particular communication of interest is itself quite difficult. A high-bandwidth line (e.g., a long-haul fiber-optic cable) typically carries hundreds or thousands of different communications; any given message may be broken into distinct packets and intermingled with other packets from other contemporaneously operating applications. The traffic on the line may be encrypted in bulk by the line provider, thus providing an additional layer of protection against the interceptor. Moreover, since a message traveling from point A to point B may well be broken into packets that traverse different physical paths en route, an interceptor at any given point in between A and B may not even see all of the packets pass by."

58. See Titanic Meeting Stuck at Dock. *Wired News.* (June 10, 1998). Of the availability of complex encryption codes outside of the United States, Microsoft CEO Bill Gates said "That's a change in the world of spying and law enforcement that we cannot effect"— meaning, precisely, that the clock cannot be turned back on encryption technologies. Likewise, the U.S. National Research Council's report, "Cryptography's Role in Securing the Information Society," concluded that "Because cryptography is an important tool for protecting information and because it is very difficult for governments to control, the committee believes that the widespread nongovernment use of cryptography in the United States and abroad is inevitable in the long run."

59. See Group of Companies to Lobby Globally on Internet Concerns. *Wall Street Journal.* (December 11, 1996).

60. The weak-strong state continuum does not capture well the transformations that are discussed here. Rather than a weakening of the state per se, what I am describing here is, rather, a re-orientation of states. (cf. Migdal 1988).

61. For a similar view on the consequences for world order of new information technologies, see James Rosenau's contribution to this volume.

REFERENCES

Agne, Philip E. and Marc Rotenberg, eds. 1998. *Technology and Privacy: The New Landscape.* Cambridge, MA: The MIT Press.

Anderson, Benedict. 1983. *Imagined Communities: Reflection on the Origins and Spread of Nationalism.* London: Verso.

Angelides, Marios C. 1997. Implementing the Internet for Business: A Global Marketing Opportunity. *International Journal of Information Management.* 17(6): 405–419.

Arquila, John, and David Ronfeldt, eds. 1997. *In Athena's Camp: Preparing for Conflict in the Information Age.* Santa Monica, CA: RAND.

Bennet, Colin J., and Rebecca Grant, eds. 1999. *Visions of Privacy: Policy Choices for the Dig ital Age.* Toronto: University of Toronto Press.

Berko, Lili. 1992. Surveying the Surveyed: Video, Space and Subjectivity. *Quarterly Review of Film and Video.* 14(1–2):61-91.

Brin, David. 1998. *The Transparent Society.* New York: Addison-Wesley.

Campbell-Kelly, Martin, and William Aspry. 1996. *Computer: A History of the Information Machine.* New York: Basic Books.

Castells, Manuel. 1989. *The Informational City: Information Technology, Economic Restructuring, and the Urban-Regional Process.* Oxford: Blackwell.

———. 1996. *The Information Age: Society, Economy, and Culture,* Vols. 1-3. Oxford: Blackwell.

Cerny, Phillip. 1995. Globalization and the Changing Logic of Collective Action. *International Organization.* 49:595–625.

Chilton, Paul. 1995. *Security Metaphors: Cold War Discourse from Containment to Common House.* New York: Peter Lang.

Computer Science and Telecommunications Board. 1996. *Cryptography's Role in Securing the Information Society.* Washington: National Research Council.

Cox, Robert. 1986. Social Forces, State and World Order." In Robert Keohane, ed. *Neo-Realism and Its Critics.* New York: Columbia University Press.

Dam, Keneth, and Herbert Lin, eds. 1996. *Cryptography's Role in Securing the Information Society.* Washington: National Research Council.

Deibert, Ronald J. 1997. *Parchment, Printing and Hypermedia: Communication in World Order Transformation.* New York: Columbia University Press.

Der Derian, James. 1996. Global Swarming, Virtual Security, and Bosnia. *The Washington Quarterly.* 19(3):45-56.

Deudney, Daniel. 1995. The Philadelphian System: Sovereignty, Arms Control and Balance of Power in the Americ States-Union, ca 1787–1861. *International Organization.* 49: 191–228.

Froomkin, A. Michael, 1997. The Internet as a Source of Regulatory Arbitrage. In Brian Kahin and Charles Nesson, eds., *Borders in Cyberspace: Information Policy and Global Information Infrastructure.* Cambridge: The MIT Press.

Gandy, Oscar Jr. 1993. *The Panoptic Sort: A Political Economy of Personal Information.* Boulder: Westview Press.

Government of Canada. 1998. *A Cryptography Policy Framework for Electronic Commerce*. Ottawa: Task Force on Electronic Commerce.

Hafner, Katie, and Matthew Lyon. 1996. *Where Wizards Stay Up Late: The Origins of the Internet*. New York: Simon and Schuster.

Hart, Jeffrey, Robert R. Reed, and Francois Bar. 1992. The Building of the Internet. *Telecommunications Policy*. (November) 16: 666–689.

Held, David. 1987. *Models of Democracy*. Stanford: Stanford University Press.

Helleiner, Eric. 1998. Electronic Money: A Challenge to the Sovereign State? *Journal of International Affairs*. 51(2): 387–409.

Herz, John. 1957. Rise and Demise of Territorial States. *World Politics*. 9: 473–493.

Huysmans, Jef. 1998. Security! What do you mean? From concept to thick signifier. *European Journal of International Relations*. 4(2): 226–255.

Innis, Harold. 1952. *The Bias of Communications*. Toronto: University of Toronto Press.

Krause, Keith, and Michael Williams. 1996. *Critical Security Studies*. Minneapolis: University of Minnesota Press.

Liphshutz, Ronnie. 1995. *On Security*. New York: Columbia University Press.

Luke, Timothy. 1991. The Discipline of Security Studies and the Codes of Containment. *Alternatives*. 16: 320–344

Lyon, David. 1994. *The Electronic Eye: The Rise of Surveillance Society*. Minneapolis: University of Minnesota Press.

Martin, David. 1991. *Geographic Information Systems and their Socioeconomic Applications*. London: Routledge Press.

McLuhan, Marshal. 1964. *Understanding Media: The Extensions of Man*. New York: McGraw-Hill.

Migdal, Joel. 1988. *Strong Societies and Weak States: State-Society Relations and State Capabilities in the Third World*. Princeton: Princeton University Press.

Strassman, Paul A., and William Marlow. 1996. Risk-Free Access Into the Global Information Infrastructure Via Anonymous Re-Mailers. Paper delivered to the Symposium on the Global Information Infrastructure: Information, Policy, and International Initiatives. Harvard University, Cambridge Mass, USA. (January 28–30)

Swett, Charles. July 1995. Strategic Assessment: the Internet. Office of the Assistant Secretary of Defense for Special Operations and Low-Intensity Conflict (Policy Planning).

Williams, Michael. 1998. Identity and the Politics of Security. *European Journal of International Relations*. 4(2): 204–255.

CHAPTER SIX

THE GLOBAL POLITICAL ECONOMY OF WINTELISM: A NEW MODE OF POWER AND GOVERNANCE IN THE GLOBAL COMPUTER INDUSTRY

SANGBAE KIM AND JEFFREY A. HART

INTRODUCTION

Technological competition in the global information industries—the leading sector in the contemporary global political economy—is currently moving beyond competition over technological innovation per se. The technological winner is now the one who manages to control de facto market standards while at the same time protecting intellectual property rights. Moreover, the new mode of technological competition puts pressure on firms and governments everywhere not only to adjust to the new principles of competition, but also to adopt new forms of industrial governance and state-societal arrangements.

In the global personal computer (PC) industry, two American companies, Microsoft and Intel, typify this new mode of technological competition. Together, Microsoft and Intel have defined the architecture for IBM-compatible PCs by setting and controlling de facto market standards and protecting those standards as the world's most valuable form of intellectual property. Scholars in International Political Economy (IPE) understand that the resurgence of the U.S. international competitiveness is closely related to its relative strength in this new leading sector. This is in a sharp contrast to the debates of the 1980s and early 1990s over the relative decline of the U.S. international competitiveness in previous leading sectors—steel, autos, consumer electronics, and semiconductors.

Building on Borrus and Zysman's work (1997), we attempt to understand the new mode of technological competition and subsequent changes in industrial governance and state-societal arrangements by using the concept of *Wintelism*. *Wintelism writ small* is a new mode of competition mainly in the personal computer industry, in which the Wintel (Windows + Intel) coalition represents the combined power of Microsoft and Intel over the architectural standards of PCs. In the PC industry, Microsoft's operating system and Intel's microprocessors are not just superior pieces of equipment that the competition might hope to match or surpass with a reasonable effort. Rather, for some years now, they have served as structural constraints—*the rules of the game*—that every firm entering the industry has had to accept.

Wintelism writ large is a new form of industrial governance that originated from the computer industry, but can be applied to all information industries. It is our view that there is a close fit between Wintelism writ large and *horizontal industrial governance*. In the Wintelist era, large firms that are vertically integrated no longer dominate because they cannot compete adequately with horizontally focused, specialized firms. We will be arguing below that recent changes in U.S. state-societal arrangements are well suited to an era of architectural competition.[1] We use the term *modified regulatory state* to refer to U.S. government policies and institutional arrangements. Other countries have not been so fortunate in this regard, including the country that was the main source of foreign competition for U.S. high technology firms in the 1980s, i.e., Japan.

THE TECHNOLOGICAL BASIS OF WINTELISM

The rise of Wintelism is connected with the growing prominence of a technological sector that we call *software electronics* technology. Software electronics technology includes computer software, microcode, semiconductor chip designs, and technical standards in products and services. Software electronics does not include the hardware aspects of electronics or information technologies. We will call these excluded technologies *hardware electronics*. Although both hardware and software electronics belong to the broader category of information technology, our definition of Wintelism begins with the distinction between the two technological sectors.

COMPUTER ARCHITECTURE TECHNOLOGIES

Among software electronics technologies, we will focus on technologies associated with *computer architecture*. Computer technology is comprised of hardware (all the physical equipment of computers), firmware (embedded software in programmable microchips) and software (a set of instructions that tells the electronics system how to perform tasks). There are also published and unpublished standards and interface protocols that allow designers to make sure that hardware and software work together. As Morris and Ferguson hold,

> The standards define how programs and commands will work and how data will move around the system—the communication protocols and formats that hardware components must adhere to, the rules for exchanging signals between applications software and the operating system, the processor's command structure, the allowable font descriptions for a printer, and so forth. (Morris and Ferguson 1993, 88)

Morris and Ferguson call this complex of standards and rules *architecture*. The architecture is mainly defined by microprocessor, basic input output system (BIOS), data bus, and operating system software. All elements are usually referred to together as a *platform*. Technologies concerning the computer architecture are the core of PC technology; among them, the most critical parts are microprocessors and operating system software.

Personal computer systems are generally designed around *microprocessors*, which embody most of the central processing unit of a computer within a single chip. The microprocessor chip is embedded in a printed circuit board with helper chips to form what is called a *motherboard*. The motherboard generally includes a separate chip for the BIOS, a digital clock, the data bus, and a bank of chip sockets for dynamic random access memory (DRAM). The motherboard is connected via the data bus and other input/output interfaces to PC peripherals such as the monitor, the keyboard, the floppy disk drives, the hard drives, and whatever else the customer wants to have connected. IBM-compatible PCs use Intel's x86 series of microprocessors or microprocessors designed to emulate those devices. Apple's Macintosh uses Motorola or IBM (Power PC) microprocessors.

Operating systems translate the software written in higher-level languages, like BASIC, Fortran, or C++ into machine language instructions that are understood by the computer's central processing unit. It also manages data flows into and out of the central processing unit and may also manage the way in which data is handled in data storage devices. In terms of the functional level of software, the operating system is most closely related to the hardware and to design a good one requires sophisticated knowledge of computer science, but does not require much knowledge in the application domain or real-world problems that end users confront. For *application software* to perform well, the designer must start from a good understanding of the problems that users are trying to solve.

TECHNOLOGICAL PROPERTIES OF SOFTWARE ELECTRONICS

There is a restricted meaning of technology as knowledge, and an extended meaning of technology in relation to embedded institutions (Hart and Kim 2000). In the restricted meaning, technology is technological knowledge embodied in material products. Here we note that there are at least three distinct aspects of technological knowledge in software electronics: technical standards, intellectual property and product innovation.[2]

The most prominent feature of contemporary software electronics technology is the increasing importance of standards. This feature derives from the high value placed by consumers on compatibility between interrelated technological components. For example, the PC is a modular device assembled from a series of discrete components, each of which has its own discernable production chain. Thus, the existence of a dominant technical standard provides producers and consumers with the advantages of compatibility among subsystems while products are continually refined and reconfigured. Architectural standards enable rapid innovation to take place at the component level without sacrificing compatibility at the system level.

Despite these advantages, architectural standards may also result in barriers to entry that lead to the potential for particular firms to exercise market power because of imperfect competition. In fact, the operating system and microprocessor are perfect examples of subsystem markets with high barriers to entry because of the entrenchment of the IBM-compatible PC architecture. To utilize the biggest selection of software for personal computers, consumers have had little choice but to buy a machine containing an Intel-designed chip and loaded with Microsoft's operating systems.

The second feature of software electronics technology is that there are increasing demands for protecting proprietary knowledge as intellectual property rights. One of the major technological trends that brings about those demands is the rising cost of research and development and other innovation-related activities. Investment in R&D has accelerated worldwide; and product life cycles have become shorter. In order to recoup substantial investments in R&D, a company must be able to secure its investment in technology in the form of intellectual property.

Effective protection of intellectual property, however, has become more difficult as copying of digital technology has become easier. For example, computer software programs costing huge sums to develop can be copied quickly and cheaply by unscrupulous individuals with fairly rudimentary equipment. Moreover, "the information-intensive nature of software means that its exploitation by a number of parties does not degrade its quality" (Mowery 1996: 305). Semiconductor chips pose similar problems with respect to existing forms of intellectual property protection. As in computer software development, designing new chips and preparing masks for chip manufacturing is expensive, but copying chip designs and reproducing chip masks is relatively simple and inexpensive.

Therefore, it is no surprise that firms want more secure ways of protecting their intellectual property. Recent evidence shows that information technology firms are seeking greater intellectual property protection through legal mechanisms, such as patents and copyrights. Similarly, the number of patent and copy-right infringement lawsuits is increasing. These current trends have raised the salience of intellectual property laws and their enforcement in the eyes of national governments (Clapes 1993; Moore 1997).

The final feature of computer architectural technologies is its unique pattern of product innovation. For computer architectures, functionality is more highly valued than quality. Software engineering is a highly knowledge-intensive process, more like a craft industry than like high-volume manufacturing. Japanese efforts to create software factories—by adapting methods from high-volume manufacturing that normally enhance productivity such as statistical quality control, standardized components, and speeded up assembly lines—have not succeeded. Instead, the most highly valued software is written by small teams of skilled engineers (Cusumano 1991).

New computer architectures are usually introduced by discontinuous break-through-type innovations rather than by incremental innovations or quality improvement. The development of a new generation of microprocessors and operating system software yields major innovations with little room for incremental improvement between major breakthroughs. Furthermore, this development of new products in microprocessors and operating systems is based on the obsolescence of old products; new products destroy the old generation (Kenney 1996).

One important question here is whether there is a set of identifiable institutional arrangements specific to a given technology or set of technologies that produces better long-term economic consequences overall for the political unit in which those institutions exist. Herbert Kitschelt (1991) argues that any technology has two important dimensions that influence the choice of industrial governance structures: one is the degree of coupling in the elements of a technological system, and the other is the complexity of causal *interactions* among production stages. He argues that each technological system—characterized by its position vis-à-vis the two dimensions—requires a distinct governance structure for maximum performance. For example, the more tightly technological elements are coupled, the more control needs to be centralized. The more complex the causal interaction between production stages is, the less control needs to be decentralized. Following Kitschelt, we argue that software electronics technology requires a distinct governance structure—or a particular set of institutional arrangement—for maximum economic performance (Hart and Kim 1998).

Software electronics is a loosely coupled technological system. Each step or component of production in a software electronics system is separated from every other step in space and time. Thus the production steps can be done in any sequence at any location because loose coupling permits decentralized control, and errors in components do not easily affect the entire system. For example, the modularity of the PC system means that parts, subassemblies, components, and peripherals can be sourced in the open market from wherever the best price/performance can be garnered. The components from multiple vendors fit together because they are compatible enough to enable end-to-end interoperability among the components. Here, architectural standards serve as the lubricant that allows modular components to work together well.

Software electronics is also a complex interactive technological system. In other words, a software electronics system requires complex feedback between production stages to keep the whole process on track. Thus, its developmental processes have to take place in decentralized organizational units, because a centralized system of control would be quickly overloaded. For example, the whole process of software development including design, coding, testing, and integration entails a tremendous amount of feedback and informal communication within the firm. Thus, technological trajectories of advanced software are not readily predictable in time, cost, or in final results. The development of new computer software technology is usually the result of trial-and-error research. This is called *learning by doing*. Likewise, close interaction between producers and sophisticated users is critical in the software development process. For instance, the alpha and beta testing of new software generations provides invaluable feedback to software developers on the features desired by users and helps eliminate bugs before the product is shipped. This is called *learning by using*. (Rosenberg 1982).

In this context, technological properties of software electronics require a flexible institutional environment that encourages the rise of decentralized industrial governance structures. Software electronics technologies do not reward the organized capabilities of vertically integrated private or state-owned enterprises or the interventionist role of the state where architectural standards exist. Smaller sized start-up firms with cross-regional or cross-national networks emerge as the fittest industrial governance structure but, in cases where R&D uncertainties are substantial and knowledge intensity is high, appropriate industrial governance requires the coexistence of large and small firms. In this regard, Herbert Kitschelt points out that,

> . . . corresponding governance structures [to software electronics] include mixed regulatory requirements and the exigencies of effective global marketing strategies give large corporations an advantage, unprecedented organizational decentralization nevertheless continues to prevail under the umbrella of the large corporation (1991, 474).

Large corporations with decentralized structures or horizontal intercorporate alliances among those corporations are required in order to provide necessary financial and technological supports. Appropriate state governance promotes this kind of horizontal industrial governance. The regulatory state promotes small start-up firms while enforcing antitrust laws to prevent large firms from discouraging innovation. The regulatory state thus encourages value-chain specialization in the computer industry as discussed below.

THE SUCCESS OF WINTEL AND STRUCTURAL POWER

The case of computer architecture technology typifies a new mode of competition in the global computer industry. In this new competition, the cutting edge of industrial competition lies in the establishment of de facto technical standards. Since the introduction of the IBM PC in 1981, for example, Intel and Microsoft have defined the IBM-compatible PC architecture and established that architecture as a global standard. In this section, we discuss how Intel and Microsoft were able to dominate markets of microprocessors and operating systems for PCs, and will explain what their success means by using the conceptual framework of structural power.[3]

THE WINTEL COALITION IN THE PC INDUSTRY

Since 1981, Intel had supplied leading edge microprocessors for IBM-compatible PCs. It maintained its leadership position in this market through development and continual improvement of its x86 series of microprocessors. All IBM-compatible PC manufacturers buy Intel-designed microprocessors or clones of Intel microprocessors to build machines that run DOS/Windows operating systems. About 90 percent of all PCs sold in recent years are IBM-compatible PCs. There are several producers of x86 Intel clone chips, such as Advanced Micro Devices (AMD) and Cyrix. However, Intel has been successful in limiting the market share for cloned microprocessors by taking deep price cuts when necessary, making steep production ramp ups of new generations of products, and launching aggressive legal challenges to companies that simply copy Intel designs rather than engineering their own design improvements.

In 1992, the year when Intel became the world's largest semiconductor manufacturer, it held the overwhelming majority of the market for the then state-of-the-art 32-bit microprocessors. Intel's share was 73 percent of this market ($3.18 billion) compared with Motorola's 8.5 percent ($0.38 billion), AMD's 8.0 percent ($0.35 billion), Texas Instruments' 1.9 percent ($0.06 billion), and NEC's 1.1 percent ($0.05 billion) (Fransman 1995, 169).

Microsoft's great opportunity came when IBM chose it to be the supplier of the DOS operating system for the PC in 1981. This gave Microsoft the basis for growth, but its subsequent performance has depended on frequent improvements in operating systems. Microsoft's market position in PC operating system software resembles that of a pure monopolist even more than Intel's in microprocessors. For example, Microsoft's operating systems sit on about 90 percent of the world's personal computers (Microsoft, 89 percent; Apple, 8 percent; Unix, 2 percent; IBM OS/2, 1 percent), and PC customers have almost no choice but to purchase DOS/Windows to access the many compatible software applications currently available on the market.

After succeeding in computer languages and operating systems, Microsoft invested in developing applications software. Its first major success in this area was a spreadsheet program called Excel, displacing Lotus, which until then had dominated the market with its 1–2–3 product. Microsoft then successfully created and marketed a word processor, Microsoft Word, for both the Apple Macintosh and the IBM PC, which managed to displace earlier programs like Word Perfect as the market leader. Microsoft now controls 60 percent of the Windows spreadsheet market, and 47 percent of the Windows word processing market. Microsoft's revenues from applications software rose from $1.4 billion in 1992 to $2.2 billion in 1993, an increase of 58 percent. Microsoft was the world's largest independent software producer in 1998, with annual revenues of $14.5 billion and 27,320 employees (Cusumano and Selby 1995, 3; Chang 1994: 15–16; http://www.microsoft.com/presspass/fastfacts.htm).

Much of Intel's and Microsoft's strength in the marketplace is the result of a special relationship they have developed over time with each other. In order to keep their share of their respective markets, Intel and Microsoft had to coordinate their strategies whenever a new microprocessor or a new version of the operating system was introduced to the market.

An introduction of a faster and more powerful microprocessor requires a new operating system to perform its tasks at higher speeds (or to perform new tasks) in order for the user to benefit from the improved chip. Similarly, the successful introduction of an operating system newly developed by Microsoft depends on the replacement of older machines that occurs whenever a new and faster Intel microprocessor is released. Independent software developers tend to focus their efforts on operating systems and hardware platforms that have the largest user base.

In a circular fashion, both Intel and Microsoft have benefited from the great variety of software applications and computer peripherals that have been developed to serve this installed base. The mutually reinforcing power of Intel's microprocessors and Microsoft's DOS/Windows products over the PC architecture gave rise to the idea of the *Wintel* (Windows and Intel) coalition. While there are some tensions in the Microsoft/Intel partnership, so far the two firms have managed to continue their successful collaboration.

WINTELIST STRATEGIES AND STRUCTURAL POWER

The question arising here is how the combined power of Wintel—or the success of Microsoft and Intel separately—was established in the PC architecture in the first place. Microeconomic theories can be used to explain Wintel's success within the conceptual frameworks of *network externalities, lock-in effects, dominant design* and *first-mover advantages*. Of particular interest is how first-mover advantages in industries subject to network externalities can be used by innovating firms to deter entry by potential market entrants and to lock in customers. These economic approaches, however, cannot predict or explain the international power implications of Wintelism.

In this section, we draw upon the concept of *structural power* to examine the power implications of Wintelism. We argue that there are basically two different ways of understanding power—*material power*, which confers the material capabilities to control over others in relational dimensions, and structural power, which confers the power to reconstitute the rules of the game (including the surrounding structure and even actor's identity) by which actors constrain other actors (Hart 1976, 1989; Hart and Kim 2000).

With this conceptual framework, we understand that, in the new mode of technological competition, an industrial winner should be able to establish the material base of manufacturing and technological innovations as usually understood. However, it should also be better able to manipulate the rules of the game of technological competition. Three aspects of Microsoft's and Intel's business strategies—control over technical standards, intellectual property protection and continuous product innovations—clearly show that the structural power dimension—interacting with the material power dimension—was working explicitly so that Wintel has remained at the center of the evolution of the PC business.

First of all, the success of Wintel should be understood within the context of the increasing importance of technical standards. Currently, competitive success flows to the company that manages to establish de facto market standards control over a broad, fast-moving, competitive product market (Gabel 1987, 1991; Grindley 1995).

The most typical example of standards competition is found in the success of IBM-compatible PCs in contrast with Apple's Macintosh series. Although some experts argue that the Macintosh architecture is technically superior to that of the IBM-PC, the latter has nevertheless stubbornly held on to its dominant market position. The IBM-compatible PC makers, by adopting an open standards strategy, effectively locked in the customer base and created a market with a much more diverse set of products with generally lower prices than comparable Apple products (Yoffie ed. 1997; Grindley 1995).

The Wintel coalition established de facto standards for each successive generation of the PC architecture by maintaining a subtle balance between aggressive diffusion and limited licensing of architectural standards—by adopting an open but owned standards strategy (Borrus and Zysman 1997). Open standards may mean a loss of market share for developers of new technology as a result of the ability of other firms to market compatible cloned products. By means of limited dissemination of standards, however, Intel and Microsoft were able to maximize the effect of network externalities and gain competitive leverage in the PC industry. Cloning may even give a systematic advantage to the initial developer if it helps it to maintain its status as first mover in successive product generations.

The competition between IBM and the clone makers led to an explosion in demand for IBM-compatible PCs. The more IBM PC clone makers used Intel chips and Microsoft operating systems and the more software developers developed products that were compatible with Wintel standards, the greater was Intel's and Microsoft's competitive advantage over potential rivals in the microprocessor and operating systems businesses. Indeed, Intel's microprocessors and Microsoft's operating systems represent a structural constraint that every firm entering the industry has had to accept in the PC business. In short, Wintel has controlled the rules of the game in the PC industry.

Theoretical works in international political economy can help us better understand the importance of technical standards for the international system as a whole. Susan Strange's concept of structural power is particularly useful. Strange argues, "structural power . . . confers the power to decide how things shall be done, the power to shape frameworks within which states relate to each other, relate to people, or relate to corporate enterprises. The relative power of each party in a relationship is more, or less, if one party is also determining the surrounding structure of the relationship" (Strange 1988: 25). According to Strange, structural power in the arena of knowledge is the most important among the four main structural arenas—security, trade, finance, and knowledge. She argues,

> . . . whoever is able to develop or acquire and to deny the access of others to a kind of knowledge respected and sought by others; and whoever can control the channels by which it is communicated to those given access to it, will exercise a very special kind of structural power. . . . today the knowledge most sought after the acquisition of relational power and to reinforce other kinds of structural power (i.e. in security matters, in production and in finance) is technology (1988, 31).

In spite of Strange's silence about technical standards, control over technical standards in the PC industry clearly qualifies as an example of structural power.

The sophisticated management of intellectual property was the second essential ingredient in the success of Wintel. While the diffusion of technical standards was a part of the offensive dimension of Wintelist strategies, the protection
of intellectual property was part of its defensive dimension. Indeed, a fine balance
between liberal dissemination of open standards and stringent protection of intellectual property rights emerged as a central issue for Intel and Microsoft.
Choosing the right degree of openness and the right amount of intellectual property rights enforcement was a problem that had to be solved, as it was, with open
but owned standards (Borrus and Zysman 1997).

To protect their interests, Intel and Microsoft were active in policing in-
fringements and taking legal action in their home markets. Interfirm level lawsuits against alleged cases of intellectual property infringement have played an
important role in protecting the intellectual property of Wintel. Examples include
computer-related infringement lawsuits, such as *NEC vs Intel, Intel vs AMD, Intel vs
Cyrix* and *Microsoft vs Shuuwa* (Clapes 1993). Without such a defense, the two
firms would not have been able to remain leaders in their respective markets, because they would not have been able to afford investments in new technologies
and new production facilities.

Successful intellectual property rights enforcement ultimately requires a commitment on the part of national governments to enact strong copyright and patent
laws in the first place and then to develop credible enforcement procedures. Intellectual property protection has primarily been a matter of national (territorial) jurisdiction in the sense that "each national government determines the scope of
protection and rights subject only to bilateral and multilateral agreements. . . .
Within each system, countries established regimes of protection that were economically and philosophically compatible with their cultures" (Hansen 1997
265-6). In recent years, Intel and Microsoft have lobbied the U.S. government for
stronger intellectual property laws, and persuaded the U.S. government to pressure foreign governments to enforce intellectual property rights. In this way, the
developers of the Wintel PC platform have been able to maintain control over
those technologies by restricting access to companies unwilling to pay the price
and follow the dictates of technology licensing agreements (Gabel 1991 11-4).

Indeed, the U.S. government has played an indirect but important role in the
international success of the Wintelist firms by advocating a strengthened international regime for protecting intellectual property. The U.S. government has taken
a trade-oriented approach to international intellectual property issues. Intellectual property protection became a major trade issue in the Uruguay Round of the
GATT and later in the World Trade Organization (WTO) Agreement on Trade-
Related Aspects of Intellectual Property Rights (TRIPs) (Ryan 1998).

However, because of the vagueness of both domestic and international intellectual property regimes, many bilateral disputes over intellectual property have

occurred in recent years. For this reason, the U.S. government has concluded a number of bilateral reciprocity agreements with other countries that protect software programming and chip mask works on substantially the same basis as in U.S. domestic law (Leaffer 1991; West 1995). Such U.S. initiatives, because of their essentially unilateral nature and the claims of target governments that the United States is violating their sovereign right to decide for themselves what intellectual property laws to enact and enforce, lead inevitably to clashes between systems—*system friction* as Sylvia Ostry calls it (Ostry 1996; Bergsten and Noland 1993; Tyson 1992).

A combination of economic and power political theories helps us understand the power implications of intellectual property protection. Intellectual property disputes have sometimes become the basis for power struggles between national governments. For example, the U.S. government has been pressuring the government of the People's Republic of China to adopt stricter intellectual property laws and to enforce them. The U.S. government has tried to persuade the Chinese government to change its legal regime for reasons of Chinese self-interest (e.g., to promote the growth of indigenous software firms), but it also has used coercion to the extent that it made stricter enforcement a condition for continuing most favored nation (MFN) trade status and U.S. support for Chinese entry into the WTO.

Beyond the relational power dimension, however, we would like to call attention to a deeper structural dimension of intellectual property issues (what Stephen Krasner calls "meta-power issues"). According to Krasner, ". . . relational power behavior refers to efforts to maximize values within a given set of institutional structures; meta-power behavior refers to efforts to change the institutions themselves . . . [and] . . . the ability to change the rules of the game" (1985: 14). International intellectual property regimes are not a given but rather must be periodically redefined by the actors themselves, while interpreting their material interests and circumstances. In her recent work, Susan K. Sell pays attention to the role of ideas—in relation to power—in helping actors to define their material interests within intellectual property regimes (1998).

In a similar vein, Joseph S. Nye's concept of *soft power* also provides a useful framework for understanding the structural dimension of intellectual property disputes (1990). Soft power is the ability to achieve desired outcomes in international affairs through attraction rather than coercion. It works by convincing others to follow, or getting them to agree to, norms and institutions that produce a desired behavior. Soft power can rest on the appeal of ideas themselves or on the ability of certain actors to set the agenda in ways that shape the preferences of others. In a related vein, Susan Strange argues that, "technological changes do not necessarily change power structures. They do so only if accompanied by changes in the basic belief systems which underpin or support the political and economic arrangements acceptable to society" (1988, 123). International intellectual property regimes are concerned with protecting the ability of individuals and private firms to influence the values and beliefs of others by means of ideas rather than tangible products—that is to say *ideational power*—and therefore must be a component of any discussion of either soft power or structural power in the contemporary international system.

Along with standards initiatives and intellectual property protection, the success of Wintel has also been based on its ability to introduce successive product innovations—the third ingredient of Wintel's success. Architectural leaders can perpetuate their market positions only if they continually offer new improved products that are compatible with older subsystems. In fact, Intel and Microsoft continuously renew their products in order to sustain their control over the PC industry. Intel's ability to maintain its leadership in the microprocessor industry lies in its ability to maintain continuity across successive product generation while at the same time greatly increasing the processing speed of its microprocessors. Microsoft has also frequently made incremental improvements and occasionally introduces major advances in its products.

A fundamental root cause of this dramatic speeding in product change has been the astonishing rate of improvement in the performance of semiconductors and software. For example, according to *Moore's Law*, the capacity of microprocessors and memory devices doubles roughly every eighteen months while the price per operation stays the same (Moore 1996; Schaller 1997). As a result of this steady and rapid technical progress, a seemingly endless stream of new personal computers is constantly being introduced. Each new introduction seems to bring greater functionality for roughly the equivalent prices of its predecessor, while the value of earlier models drops dramatically. As Martin Kenney argues, "as value is being created more quickly, it is also being destroyed more quickly . . . the economy is obsolescence-based" (Kenney 1996). The rapid obsolescence of successful knowledge-intensive products, often accelerated by the new products developed and introduced to the market by the original innovators, is often cited as a contemporary example of Schumpeterian *creative destruction* (Schumpeter 1950).

Microsoft and Intel have managed to subtly balance the three strategies of being aggressive in diffusing standards, innovative in periodic improvement of products, and fiercely protective of intellectual property rights. There is an unavoidable trade-off among these three strategies, since, for example, too aggressive protection of intellectual property can result in slower diffusion of standards. We will argue below that one important role for public policy is to make sure that aggressive protection of intellectual property does not become an impediment to market growth or market entry by potential competitors.

The dynamics of technological competition are not determined solely by the actors' material capabilities in production and innovation. The competition is as much about structural power as it is about material power. This new form of structural power also changes the nature of competition over material resources by privileging certain types of technological knowledge in the broader international competition for resources.

Stefano Guzzini's recent work on power, which implicitly assumes a constructivist stance, provides a useful framework for synthesizing our discussion of material power vs. structural power.[4] Guzzini focuses on the interaction between *agent power* and *structural governance*, saying that "power lies both in the relational interaction of agents and in the systematic rule that results from the consequences of their actions. . . . power analysis, as the comprehensive account of power phenomena, must call into question the relationship between the different forms of power and of governance" (1993, 471–4).

Obtaining control over key material resources is a primal reason for exercising power. Any cultural, institutional, or normative developments that improve the efficacy of those resources should logically also become targets for attempts to acquire power. As Guzzini suggests, therefore, "two strategies are possible to improve one's potential power in a given situation: to cause either a quantitative improvement of the relevant situational power resources or a change in the environment that defines the situationally relevant power resources" (1993, 455–6). The question arising here is how to improve both resources and environment—material power as well as structural power in our terms—in a specific issue area, for example the global computer industry. To answer this question, we turn now to the role of industrial and governmental institutions.

THE RISE OF WINTELISM AND DECENTRALIZED GOVERNANCE

The rise of architectural competition pressures firms and governments everywhere not only to adjust to the new principles of competition, but also to reconsider their institutional environment to adjust better to technological and competitive changes. We will argue that modifications in American institutions in response to increased competition from other industrialized countries (especially Japan) helped to assure the success of Wintelism and the resurgence of the U.S. industrial competitiveness. In this section, we focus on two levels of governance: industrial governance structures and state-societal arrangements.

HORIZONTAL INDUSTRIAL GOVERNANCE

Industrial governance is defined mainly in terms of corporate and industry structure. Types of corporate governance can be distinguished by observing the characteristics of firms and industries: for example, whether coordinating networks are

organized vertically or horizontally. Those characteristics include the size of the industry, the organizational structure of firms, the degree of concentration of ownership, the level of inter-firm coordination, the degree to which user-producer (or
manufacturer-supplier) links are utilized by firms in the industry, and the presence
of national or cross-national production and distribution networks.

The pattern of integration among industrial units in the U.S. computer industry is horizontal. The so-called Silicon Valley model typifies the horizontal
governance structure (Ferguson and Morris 1994). The Silicon Valley model encourages horizontally focused and non-bureaucratic corporate structures. In ar-
chitecturally contending companies like Microsoft, for example, architectural
competition permits many systems and organizations to be developed independently and still work together gracefully. It also permits clean separation between
centralized general-purpose functions and decentralized or specialized functions,
and enables management of unpredictability and change (Ferguson and Morris
1994; Cusumano and Selby 1995; Cusumano and Smith 1997).

The competitive structure of the U.S. computer industry is also being transformed from one of an oligopoly dominated by large vertically integrated firms to
something else. The PC industry from its earliest beginnings adopted a horizontal
supplier structure, consisting of competing PC assembler firms. Companies such
as Intel, Microsoft, Novell, Lotus, Compaq, Seagate, Oracle, 3Com, Electronic
Data Systems, and many others all thrived by being specialists in particular layers
of a newly emerging information technology industry value chain. All these firms
were integrated in a horizontal way—not vertically as the older mainframe computer companies were—and formed regional production networks in Silicon Valley (Borrus and Zysman 1997; Cringely 1993; Saxenian 1994).

In the horizontal industrial structure, a handful of companies supplying
components to PC assemblers came to define and control the system's critical architectures, each for a specific layer of the system. For example, Borrus and Zysman hold,

> Market power has shifted from the assemblers—such as Compaq, Gate-
> way, IBM, or Toshiba—to key producers of components (e.g., Intel); oper-
> ating systems (e.g., Microsoft); applications (e.g., SAP, Adobe); interfaces
> (e.g., Netscape); languages (e.g., Sun with Java); and to pure product defi-
> nition companies like Cisco Systems and 3COM (1997, 150).

This shift in market power is suggested in the advertisements of PC producers like
IBM, Toshiba, Compaq or Siemens-Nixdorf. Their systems are nearly identical
and emphasize components or software that have become de facto market standards—*Intel Inside* and *Microsoft Windows Installed*—rather than unique features of
their own brands.

Another important point is that the horizontal supplier network in the electronics industry reaches to the global arena. To describe the global production networks, Borrus and Zysman (1997) adopt the concept of *Cross National Production Networks* (CPNs). CPNs refer to the disintegration of the industry's value chain into constituent functions that can be contracted out to independent producers wherever those companies are located in the global economy. CPNs now affect the entire global electronics industry. Moreover, CPNs express the reduced need for companies to control production through ownership or direct management of each piece of the value chain.

Indeed, Wintelist strategies of relying on product standards control and intellectual property protection facilitate the rise of global crossnational production networks. A given firm can more easily subcontract production, even across national boundaries, without worrying about the possibility that contract suppliers will develop competitive technologies because that firm can still dominate the market for critical systems elements through setting de facto market standards. Wintelism creates a whole range of market opportunities for de facto standards holders in sectors that were previously dominated by giant assemblers.

THE MODIFIED REGULATORY STATE

State governance is mainly defined by the industrial role of the state. The so-called strength of the state—the capabilities of government agencies and other national political institutions in relation to the business sector, including mechanisms of state penetration into society—or state-societal arrangements—defined in terms of the distribution of power among the state, the private business sector, and organized labor—is often considered to be a critical factor for understanding the nature of state governance (Hart 1992). More specifically, the industrial role of the state is embodied as industrial policy, which refers to the deliberate attempt by the government through a range of specific policies such as financial subsidies, trade protectionism, promotion of R&D, and procurement to determine the structure of the economy (Johnson 1982).

The U.S. state is often considered to be a regulatory state; it is frequently contrasted with the developmental states of East Asia—particularly of Japan and South Korea—that intervene directly in industrial matters and try to direct investment into high priority areas.[5] Nevertheless, the U.S. government has intervened in certain industries where there is a clear national security or public goods rationale. For example, as an advanced user and R&D sponsor, the U.S. government made important contributions throughout much of the history of computer industry—but particularly in the early period.

The so-called first-mover advantages of the American computer industry were generated not only as a result of commercial activity but also by government R&D policies, often through the Defense Advanced Research Projects Agency (DARPA) (Flamm 1987, 1988). The defense-oriented (or mission-oriented) R&D policy of the postwar U.S. government, which included substantial public funding of basic computer science research in universities, was very important in establishing the technological basis for the computer industry (Ergas 1987). The technologies created in some military programs were spun off into commercial computers; this largely unplanned diffusion and sharing of technology resulted in first-mover advantages for the American computer industry. IBM's entry into electronic computers, for example, was largely underwritten by military contracts (Alic 1992; Sandholtz et al. 1992).

However, in order to understand the evolution of U.S. government policies toward the computer industry, we need to look beyond the boundaries of what is considered industrial policy—i.e., industrial targeting, subsidies and R&D programs. Industrial policy is designed to help specific industries to achieve and/or maintain global competitiveness. The regulatory pattern of government policy in the U.S. computer industry, which relied on macroeconomic policies, antitrust enforcement, and vigorous Intellectual Property Rights protection, was more important for the growth of that industry than any industrial policy. In particular, we should note that there are two types of important regulatory government policies for the PC industry.

The U.S. government, by strictly enforcing antitrust and fair competition laws, made important, but often largely unrecognized, contributions to the rise of Wintelism. U.S. enforcement of antitrust and fair trading laws in the 1960s led to IBM's unbundling of hardware and software sales, which was central in encouraging value-chain specialization in the computer industry and fostering the growth of both the semiconductor and packaged software industries (Mowery 1994; Mowery ed. 1996). Indeed, the policy-induced emergence of computer component suppliers began subtly to undermine the logic of competition rooted in economies of scale and vertical control of technology. They helped to create the foundation for the emergence of Wintelism.

During the 1980s and early 1990s, the United States government seemed to be relaxing its antitrust policies, especially in sectors with strong R&D and strong foreign competition, and owners of intellectual property rights benefited from a more benign judicial attitude (Merges 1996). Recent actions against Intel and Microsoft taken by the U.S. Department of Justice and the Federal Trade Commission suggest a revival of interest in stricter enforcement of antitrust and fair trading laws.

The U.S. government has also played a major role in promoting Wintelism by defining and protecting the intellectual property rights of major firms. The U.S.

government has long recognized the importance of protecting intellectual property in industry as a way of encouraging technological innovation. Intellectual property is seen as a key asset for modern corporations with very important ramifications for industrial strategy and structure. Merges holds,

> Intellectual property determines the degree of legal shelter an incumbent can count on. Strong protection, like a brick wall, protects such an incumbent from the winds whipped up by potential entrants, while weak protection is more like a tent—it helps but cannot be relied on when the winds get too strong (1996: 285).

The legal development of computer program-related intellectual property laws suggests that the United States has adopted a strong protection regime for computer hardware as well as software.

GOVERNANCE STRUCTURES IN SOFTWARE ELECTRONICS

Success or failure in software electronics basically depends on the match between the architectural technological competition of the industry and national institutional arrangements, as argued at the beginning of this essay. The technological properties of software electronics—a loosely coupled system with high causal complexity—are consistent with the rise of horizontal industrial governance, as seen in the U.S. PC industry and in the Silicon Valley model. The properties of software electronics are also consistent with a modified regulatory relationship between the government and other economic actors.

Industrial governance structures in the U.S. computer and software industry approximate a flexible form that blends competition and cooperation in order to cope with the unique innovative patterns of software electronics: large R&D costs, trial and error research yielding fast-paced breakthrough-type innovations, and inexpensive copying and distribution of digital media. Small venture capitalists invest in those nodes of the innovation network in which causal relations are sufficiently well understood. Also, rapid innovative patterns in software electronics is likely to give large corporations with decentralized structures an advantage over more centralized research arrangements—like cooperative R&D consortia—in developing new products and in bringing them to the marketplace.

In cases where R&D uncertainties are substantial, a comprehensive *public and semipublic* infrastructure of technological development through universities and public research centers plays a critical role. For example, the defense-oriented industrial policy of the postwar U.S. government, which included substantial public funding of basic computer science research in universities—without any clear

industrial applications—has been more successful in establishing a strong domestic industry than interventionist forms of industrial policy in other countries.
Moreover, the regulatory role of the U.S. government to provide a competitive
market situation has been working as a political foundation for the success of the
industry.

This characterization of governance structures in the U.S. computer industry
parallels in important ways the general features of the American system of political economy. Robert Gilpin argues,

> Corporate governance in the United States is characterized by extensive
> fragmentation and an overall lack of policy coordination at both the na-
> tional and, to a lesser extent, the firm level. As in the case of the gov-
> ernment, a primary motive behind this fragmentation of corporate
> organization is to prevent the concentration of power. . . . the American
> system fits the neoclassical model of a pure competitive model based on
> price competition and in which firms seek to maximize profits (1996,
> 419–20).

In cases where countries already have elements of appropriate governance structures fitted into technological properties of software electronics within their existing national institutions, there are more possibilities that technological success
will be achieved within a framework of *path-dependent learning* (Hart and Kim
1998). In this sense, the U.S. computer industry benefited enormously from existing governance structures conducive to software electronics innovations. The
U.S. case shows that, when following the process of path-dependent learning, the
initial costs of entering new electronics technology markets are quite modest and
therefore even relatively small firms will be able to respond to new opportunities
quickly.

The American science and technology (S&T) infrastructure may be uniquely
well suited for architectural competitions in new technologies. According to Margaret Sharp, S&T infrastructure involves high quality secondary education, a
good vocational training system, a strong university sector, a well-funded academic research base with a major postgraduate component, university-industry
linkage, research associations that support technology dissemination to small and
medium-sized businesses, and the encouragement of regional initiatives bringing
together firms, universities and research institutions (1997, 101). Indeed, the role
of a social, cultural or institutional infrastructure in producing human resources
and technological knowledge gains attention especially in software electronics.

The domestic system of higher education in the United States, for example,
appears to provide a much thicker basis of appropriate human resources for software electronics than those in Japan or Europe. The structure of American higher

education systems also has closer links with government-funded research in the computer sector. American Universities have maintained closer relationships with corporations in producing and sharing technological knowledge. In fact, the organizational and disciplinary flexibility of U.S. universities in computer science has not been matched in many of the other economies. This S&T infrastructure has been supported by the unique American technological culture encouraging breakthrough-type and creative but risky innovative attempts in software electronics (Mowery 1996, 306–307; Nelson 1998, 321).

To conclude, the U.S. success in creating Wintelism provides a better explanation of the recent resurgence of U.S. international competitiveness. In effect, the rise of Wintelism enabled U.S. firms to pioneer the new rules of the game in the global computer industry: ones that grew out of the distinctively American market environment and were adapted to overseas opportunities. In the PC industry, for example, U.S. firms lead the industry overall and also dominate many segments including complete systems, microprocessors, operating systems, and packaged applications (Dedrick and Kraemer 1998, 58). United States firms were able to set global standards because they not only had the ability to maintain and expand their spheres of control, but also were supported by (and adjusted to) the American system of political economy.

CONCLUSIONS

The development of software electronics gave rise to a new mode of technological competition where control over architectural standards became more important than advanced manufacturing capabilities. We have tried in this essay to argue that the success of the Wintel coalition—Microsoft and Intel—in the global PC market is an indication of the rise of a new mode of technological competition called Wintelism that is much broader in scope (Wintelism writ large). We claimed that an assessment of the political implications of Wintelism requires a definition of power that goes beyond the conventional understanding of power in terms of control over material resources. Wintelist firms have concentrated on a set of strategies—creative use of technical standards and intellectual property rights, backed by accelerated innovations—that enabled them to define the rules of the game in horizontal markets. They became hegemonic in their horizontal niches. The dominance of technological architectures characteristic of Wintelism is therefore a form of structural power.

This horizontal hegemony poses very interesting problems for governance at the level of the national government. Should national governments promote horizontal hegemony in the name of international competitiveness (as Bill Gates so urgently argues is necessary) or should they enforce their antitrust and competition laws and break up horizontal monopolies to prevent predatory pricing and unfair trade practices? Should they fund new R&D projects proposed by horizontal hegemons or should they preferentially fund small challengers to those hegemons?

Relying on the previous work of Herbert Kitschelt, we argued that innovations in software electronics, a loosely coupled technological system with high causal complexity, are consistent with horizontal industrial governance structures and a modified regulatory state. The regulatory state in the United States has been modified slightly to make it possible to produce a large number of high quality computer professionals and to permit close university-industry linkages for creative research. From time to time, the U.S. government condones industrial targeting. For example, the creation of R&D consortia like Sematech for the semiconductor industry was permitted as an exception to the general rule of avoiding direct interventions in industrial development.

As described above, Wintelism was born in the transition from the mainframes to PCs in the computer industry and as a response to increased competition from Japan in the 1980s; but we are now observing another transition in the computer industry. Since the early 1990s, there have been signs of the growing importance of a combined computer and telecommunications industry that increasingly revolves around global network infrastructures (Moschella 1997). As this network-centric era begins, the prospects for new market leaders and new types of power are once again topics of speculation.[6] The critical question that arises here is whether the current shift toward networked computers will result in the same kinds of fundamental changes across a wide range of customer, technology, distribution, sales, marketing, and supplier businesses that characterized the rise of Wintelism or whether Wintelism will simply adapt itself to the increased importance of network computing.

NOTES

An earlier version of this paper was prepared for delivery at a workshop on *Information, Power and Globalization* at the annual meeting of the American Political Science Association, Boston, Massachusetts, September 2 1998. We would like to express our gratitude to the following for comments and criticisms on earlier drafts: Jonathan Aronson, Sandra Braman, Ed Comor, Ron Deibert, Rob Kling, Karen Litfin, Stephen McDowell, James N. Rosenau, J.P. Singh, Debora Spar, Virginia Walsh, Mark Zacher, John Zysman, and several anonymous reviewers. Please do not cite or quote without the written permission of the authors.

   1. Please note the parallels between our argument on this subject and those of Braman, Comor and Deibert in this volume.

2. Based on this new conceptualization of technology, we can attempt to go beyond the inherited and limited view of technology as easily transferrable, proprietary knowledge. We propose a more complex concept of technology that includes technical standards, intellectual property rights, norms, craft knowledge, and embedded institutions and culture. Hart and Kim (2000) coined a new term, *technoledge*, compounded from technology and knowledge, in order to emphasize this new and more complex conceptualization of technology.

3. Concerning the architectural dominance and business strategies of Intel and Microsoft, there are many well-documented works. For example, for the rise of Intel and Microsoft in the context of IBM's collapse, see Chposky and Leonsis (1988), Carroll (1993), and Ferguson and Morris (1994). Concerning Intel and its microprocessor business, see Moore (1996) and Jackson (1997). Concerning Microsoft and Bill Gates, see Wallace and Erickson (1992), Ichbiah and Knepper (1992), Manes and Andrews (1993), Cusumano and Selby (1995), Stross (1996), and Wallace (1997).

4. International theorists are recently thinking more about the larger set of norms, rules, and structures, which have governed international systems. For example, works by IR theorists, such as Alexander E. Wendt (1987 1992), in the tradition of social constructivism, have taken their cues from Anthony Giddens's structuration theory in sociology (1984). In the most recent work in this tradition, the constructivists have made the formation of identities and social norms a key question for research (Katzenstein, ed. 1996).

5. See the essays by Aronson, McDowell, and Zacher in this volume for further discussion of the differences among the advanced industrial nations in their approaches to regulating the computer and telecommunications industries.

6. See the essays by Aronson, Singh and Zacher in this volume for further discussion of the impact of network technologies.

REFERENCES

Alic, John A. et al. 1992. *Beyond Spinoff: Military and Commercial Technologies in a Changing World.* Boston: Harvard Business School Press.

Bergsten, C. Fred, and Marcus Noland. 1993. *Reconcilable Differences?: United States-Japan Economic Conflict.* Washington D.C.: Institute for International Economics.

Borrus, Michael and John Zysman. 1997. Globalization with Borders: The Rise of Wintelism as the Future of Global Competition. *Industry and Innovation.* 4(2).

Carroll, Paul. 1993. *Big Blues: The Unmasking of IBM.* New York: Crown Trade Paperbacks.

Chang Jr., Ike Y. 1994. *The Economics of Dominant Technical Architectures: The Case of the Personal Computer Industry.* Santa Monica, CA: RAND.

Chposky, James, and Ted Leonsis. 1988. *Blue Magic: The People, Power and Politics Behind the IBM Personal Computer.* New York: Facts on File.

Clapes, Anthony L. 1993. *Softwars: The Legal Battles for Control of the Global Software Industry.* Westport, Conn.: Quorum Books.

Cringely, Robert. 1993. *Accidental Empires: How the Boys of Silicon Valley Make Their Million, Battle Foreign Competition, and Still Can't Get a Date.* Reading, MA: Addison-Wesley.

Cusumano, Michael A. 1991. *Japan's Software Factories: A Challenge to U.S. Management.* New York: Oxford University Press.

Cusumano, Michael A., and Richard W. Selby. 1995. *Microsoft Secrets: How the World's Most Powerful Software Company Creates Technology, Shapes Markets, and Manages People.* New York: Free Press.

Cusumano, Michael A., and Stanley A. Smith. 1997. Beyond the Waterfall: Software Development at Microsoft. In David B. Yoffie, ed. *Competing in the Age of Digital Convergence.* Boston: Harvard Business School Press.

Dedrick, Jason, and Kenneth L. Kraemer. 1998. *Asia's Computer Challenge: Threat or Opportunity for the United States & the World?* New York: Oxford University Press.

Drucker, Peter F. 1993. *Post-capitalist Society.* New York: HarperBusiness.

Ergas, Henry. 1987. Does Technology Policy Matter? In Bruce R. Gile and Harvey Brooks, eds. *Technology and Global Industry: Companies and Nations in the World Economy.* Washington, D.C.: National Academy Press.

Ferguson, Charles H., and Charles R. Morris. 1994. *Computer Wars: The Fall of IBM and the Future of Global Technology.* New York: Times Books, Random House.

Flamm, Kenneth. 1987. *Targeting the Computer: Government Support and International Competition.* Washington, D.C.: The Brookings Institution.

Flamm, Kenneth. 1988. *Creating the Computer: Government, Industry, and High Technology.* Washington, D.C.: The Brookings Institution.

Fransman, Martin. 1995. *Japan's Computer and Communications Industry: The Evolution of Industrial Giants and Global Competitiveness.* Oxford: Oxford University Press.

Gabel, H. Landis, ed. 1987. *Product Standardization and Competitive Strategy.* Amsterdam: North-Holland.

Gabel, H. Landis. 1991. *Competitive Strategies for Product Standards: The Strategic Use of Compatibility Standards for Competitive Advantage.* London: McGraw-Hill.

Giddens, Anthony. 1984. *The Constitution of Society: Outline of the Theory of Structuration.* Cambridge, England: Polity.

Gilder, George. 1989. *Microcosm: The Quantum Revolution in Economics and Technology.* New York: Touchstone.

Gilpin, Robert. 1996. Economic Evolution of National Systems. *International Studies Quarterly.* (September) 40(3).

Grindley, Peter. 1995. *Standards Strategy and Policy: Cases and Stories.* Oxford: Oxford University Press.

Guzzini, Stefano. 1993. Structural Power: the Limits of Neorealist Power Analysis. *International Organization.* 47(3).

Hamilton, Marci A. 1997. The TRIPs Agreement: Imperialistic, Outdated, and Overprotective. In Adam D. Moore, ed. *Intellectual Property: Moral, Legal, and International Dilemmas.* Lanham, MD: Rowman & Littlefield.

Hansen, Hugh C. 1997. International Copyright: An Unorthodox Analysis. In Adam D. Moore, ed. *Intellectual Property: Moral, Legal, and International Dilemmas.* Lanham, MD: Rowman & Littlefield.

Hart, Jeffrey A. 1976. Three Approaches to the Measurement of Power in International Relations. *International Organization.* 30(2).

——. 1989. ISDN and Power. Discussion Paper 7, Center for Global Business, the Business School of Indiana University.

——. 1992. *Rival Capitalists: International Competitiveness in the United States, Japan, and Western Europe.* Ithaca: Cornell University Press.

Hart, Jeffrey A. Hart, and Sangbae Kim. 1998. Technological Capacity as Fitness: an Evolutionary Model of Change in the International Political Economy. Paper prepared for delivery at the conference on *Evolutionary Perspectives on International Relations*, Bloomington, IN. (December 4–6)

——. 2000. Power in the Information Age. In Jose V. Ciprut, ed. *Of Fears and Foes: Security and Insecurity in an Evolving Global Political Economy.* Westport, Conn.: Praeger.

Ichbiah, Daniel, and Susan L. Knepper. 1992. *The Making of Microsoft: How Bill Gates and his Team Created the World's Most Successful Software Company.* Rocklin, CA: Prime Pub.

Jackson, Tim. 1997. *Inside Intel: Andy Grove and the Rise of the World's Most Powerful Chip Company.* New York: Dutton.

Johnson, Chalmers. 1982. *MITI and the Japanese Miracle: The Growth of Industrial Policy 1925–1975.* Stanford: Stanford University Press.

Katzenstein, Peter J., ed. 1996. *The Culture of National Security: Norms and Identity in World Politics.* New York: Columbia University Press.

Kenney, Martin. 1996. The Role of Information, Knowledge and Value in the Late 20th Century. *Working Paper.*

Kitschelt, Herbert. 1991. Industrial Governance Structures, Innovation Strategies and the Case of Japan: Sectoral or Cross-National Comparative Analysis. *International Organization.* 45(4).

Krasner, Stephen D. 1985. *Structural Conflict.* Berkeley: University of California Press.

Leaffer, Marshall A. 1991. Protecting United States Intellectual Property Abroad: Toward a New Multilateralism. *Iowa Law Review*. 76.

Manes, Stephen, and Paul Andrews. 1993. *Gates: How Microsoft's Mogul Reinvented an Industry and Made Himself the Richest Man in America*. New York: Doubleday.

Merges, Robert P. 1996. A Comparative Look at Intellectual Property Rights and the Software Industry. In David C. Mowery, ed. *The International Computer Software Industry: A Comparative Study of Industry Evolution and Structure*. New York: Oxford University Press.

Moore, Adam D., ed. 1997. *Intellectual Property: Moral, Legal, and International Dilemmas*. Lanham, MD: Rowman & Littlefield.

Moore, Gordon E. 1996. Intel: Memories and the Microprocessor. *Daedalus*. (Spring) 125(2).

Morris, Charles R., and Charles H. Ferguson. 1993. How Architecture Wins Technology Wars. *Harvard Business Review*. (March–April)

Moschella, David C. 1997. *Waves of Power: Dynamics of Global Technology Leadership, 1964–2010*. New York: AMACOM.

Mowery, David C. 1994. *Science and Technology Policy in Interdependent Economics*. Boston: Kluwer Academic Publishers.

Mowery, David C., ed. 1996. *The International Computer Software Industry: A Comparative Study of Industry Evolution and Structure*. New York: Oxford University Press.

Nelson, Richard R. 1998. The Co-evolution of Technology, Industrial Structure, and Supporting Institutions. In Giovanni Dosi, David J. Teece, and Josef Chytry, eds. *Technology, Organization, and Competitiveness: Perspectives on Industrial and Corporate Change*. Oxford: Oxford University Press.

Nye, Jr. Joseph S. 1990. *Bound to Lead: The Changing Nature of American Power*. New York: Basic Books.

Ostry, Sylvia. 1996. Policy Approaches to System Friction: Convergence Plus. In Suzanne Berger and Ronald Dore, eds. *National Diversity and Global Capitalism*. Ithaca and London: Cornell University Press.

Rosenberg, Nathan. 1982. *Inside the Black Box: Technology and Economics*. Cambridge University Press.

Ryan, Michael P. 1998. *Knowledge Diplomacy: Global Competition and the Politics of Intellectual Property*. Washington, D.C.: Brookings Institute.

Sandholtz, Wayne, Michael Borrus, John Zysman, Ken Conca, Jay Stowsky, Steven Vogel, and Steve Weber. 1992. *The Highest Stakes: The Economic Foundations of the Next Security System*. New York: Oxford University Press.

Saxenian, Annalee. 1994. *Regional Advantage: Culture and Competition in Silicon Valley and Route 128*. Cambridge: Harvard University Press.

Schaller, Robert R. 1997. Moore's Law: Past, Present, and Future. http://www.cs.uoregon. edu/classes/ cis629/Readings/moore.htm.

Schumpeter, Joseph A. 1950. *Capitalism, Socialism and Democracy*. 3rd ed. New York: Harper and Row.

Sell, Susan K. 1998. *Power and Ideas: North-South Politics of Intellectual Property and Antitrust*. Albany: State University of New York Press.

Sharp, Margaret. 1997. Technology, Globalization, and Industrial Policy. In Talalay, Michael, Chris Farrands, and Roger Tooze, eds. *Technology, Culture, and Competitiveness: Change and the World Political Economy*. London and New York: Routledge.

Strange, Susan. 1988. *State and Markets*. London: Pinter.

Stross, Randall E. 1996. *The Microsoft Way: The Real Story of How the Company Outsmarts Its Competition*. Reading, MA: Addison-Wesley.

Toffler, Alvin. 1990. *Power Shift*. New York: Bantam Books.

Tyson, Laura D'Andrea. 1992. *Who's Bashing Whom?: Trade Conflict in High-Technology Industries*. Washington, D.C.: Institute for International Economics.

Wallace, James. 1997. *Overdrive: Bill Gates and the Race to Control Cyberspace*. New York: Wiley.

Wallace, James, and Jim Erickson. 1992. *Hard Drive: Bill Gates and the Making of the Microsoft Empire*. New York: HarperBusiness.

Wendt, Alexander E. 1987. The Agent-Structure Problem in International Relations Theory. *International Organization*. 41(3).

Wendt, Alexander E. 1992. Anarchy is What States Make of It: the Social Construction of Power Politics. *International Organization*. 46.

West, Joel. 1995. Software Rights and Japan's Shift to an Information Society: The 1993–1994 Copyright Revision Process. *Asian Survey*. 35(12).

Yoffie, David B., ed. 1997. *Competing in the Age of Digital Convergence*. Boston: Harvard Business School Press.

CHAPTER SEVEN

NEW TECHNOLOGIES AND CONSUMPTION: CONTRADICTIONS IN THE EMERGING WORLD ORDER

EDWARD COMOR

Capitalism is inherently innovative. In a competitive capitalist system, the need to implement increasingly efficient ways of producing, distributing and selling commodities is ever present. But nations, communities and individuals are not machines—people can not be readily manipulated or "upgraded" in ways that are always accommodating to such systemic compulsions. New technologies, regardless of the logic of their development, are rarely embraced wholeheartedly. Indeed, sometimes they are rejected. Simply put, what is rational for capital may not be acceptable to human beings.

A remarkably under-assessed aspect of this tension is the ability or willingness of people to purchase the goods and services that global capitalism is producing. Not only are technologies directly developed for the production and distribution of commodities, they are now becoming increasingly important in efforts to further consumption. Students of International Studies have not yet directly addressed the complexities and implications of this relationship between technology and consumption and these complexities and implications constitute the focus of what follows.

Following the collapse of communism and the near universalization of neoliberal economic policies, at the beginning of the twenty-first century, the world order is in many ways a characteristically capitalist world order. While the implications of production and distribution developments and the role of new technologies in relation to these have been examined by students of International Political Economy and others, analyses on how technology is shaping consumption—the final stage in the production process—largely has been ignored. More generally, a vast range of governance developments are being directly influenced by more than just political actors pursuing various interests involving rapidly emerging technological capacities. Governance also is being shaped by technological applications being implemented to modify an essential institution of human existence—consumption. (For other chapters in this volume that address the relationship of capitalism to governance, see McDowell, and Kim and Hart).

Borrowing from economist Thorstein Veblen's conceptualization of consumption as an institution—a structured complex of "habits of thought" (Veblen 1953, 133)[1]—in the following pages, I discuss the relationship between consumption and technology with a particular emphasis on the problems involved in coordinating production with consumption. In the first section, I argue that consumption can and should be conceptualized as a sociological institution. I do this in order to establish the complex nature of consumption, both in relation to the capitalist production process and to reality in toto. In the second section, the role of technology in shaping this institution is addressed. Systemic forces are identified that compel capitalists directly and state officials indirectly to expand markets through what I call the global "widening" and "deepening" of consumption. In the section following this, I focus on the Internet as a contemporary case study of these. An argument is made here that Internet developments potentially could lead to a crisis of consumption and, hence, global capitalism writ large. This position is elaborated in the fourth and penultimate section.

As Robert W. Cox has recognized, "Consumption is the motor force of capitalism and the motivation of consumer demand is indispensable to capitalism's continuing development" (1995, 168). In the context of this importance, my goal in this chapter is to locate where and when systemic tensions and potential contradictions might occur in light of technological and related transformations in the global political economy. (For a quite different approach to issues related to global consumption, see the chapter by Aronson.)

CONSUMPTION AS AN INSTITUTION

It is useful to think of consumption as more than just a necessary activity. More generally, consumption also is an institution—a historically constructed and power-laden typification of habitualized thoughts and actions. Indeed, the very raison d'etre of all institutions is to structure ways of thinking and doing (Berger and Luckmann 1967, 54). Institutions influence all human relationships and their pervasiveness, for the most part, is a result of their functional necessity. Through this structuring of thought and action, predictability in most interactions is facilitated and this relieves people of having to spend considerable time, effort and, in some instances, the need to engage in explicit conflict in virtually every social encounter. Also, over time, experience (sometimes bitter) tends to compel people to appreciate things as they are as opposed to how they might be. This distaste for uncertainty and unpredictability further entrenches institutional norms.

Thoughts and actions involving consumption are constructed realities involving class, gender, and other power dynamics. As with most institutions, such historically produced realities directly affect thought and action in ways that deny their historicity. The way in which consumption is conceptualized and practiced in any given place and time is not only a human-produced reality, ways of thinking about consumption and "doing" consumption themselves are structured practices. This, in effect, limits the flexibility of such institutionalized ways of thinking and doing and herein lies a potential contradiction: while the inherent dynamism of capitalism compels ongoing reforms to the institution of consumption, there are limits to how fast consumption ideals and practices can in fact be modified over a given period of time in a particular spatial context.

The beginnings of modern consumerist societies usually are traced to the industrial revolution. Understanding the forces and processes that led to this involves an appreciation of already existing institutions, organizations and technologies and their mediation of consumption-related developments. Pre-capitalist grain purchasing practices, for example, mediated and shaped the development of early English capitalism. The long-standing custom of selling grain in open markets on specific days at a commonly agreed upon just price—involving militant interventions when farmers, merchants or millers were suspected of manipulating supplies to increase prices—directly shaped subsequent norms in seller-buyer relations (Ackerman 1997, 113–14). What remains alive in thought and practice directly shapes the parameters of what is possible. In conceptualizing consumption as an institution, history is a tool for analysts seeking to specify process and identify tensions and contradictions in contemporary developments.

At the beginning of the twenty-first century, efforts to stimulate consumer activity over the Internet are being directly influenced by traditional shopping practices such as the seemingly natural proclivity to physically handle products and interact with other living, breathing human beings. As with other structured relationships, a core part of the ongoing construction of the institution of consumption involves not just problematic revisions to established ways of thinking and doing but also the development and active interventions of new mediators in daily life. Through the pervasiveness of institutions such as the price system, organizations like the advertising-marketing firm, and technologies such as automatic teller machines, what has come to be known as the consumer society continues to develop in uneven and sometimes unpredictable ways.

Consumption is like other institutions in terms of its structuring proclivities but it is relatively unique in terms of its extraordinary position in capitalist economies. To better understand these dynamics and the role of technological developments in relation to them, it is helpful to recall the classical Marxist dialectic between what are called forces of production and relations of production. By forces of production, I refer to the devices (e.g. technologies), facilities (e.g. factories) and ideas (e.g. technical know-how) used in the production process. By relations of production, I refer to how human relationships are organized in accordance with ways of doing production, distribution, exchange, and consumption. At any given place and time, relations of production roughly correspond with the state of existing productive forces. For example, productive relations—as expressed through school systems, legal structures, religious orders, and so forth—to some extent are expressions of the ways in which people make use of productive forces such as natural resources, manufacturing capabilities, social pools of knowledge and, of course, technologies (Cohen 1988, 4–7).

This forces-relations dialectic continues to play itself out through, for instance, the recent flowering of computer-based information and communication technologies. Their widespread development and application can be traced to the economic crisis of the 1970s—a period in which mostly Western-based corporations and states were compelled to respond to unprecedented conditions of rising inflation, deepening unemployment, and growing competition from overseas. Measures to reorganize production norms and related reforms to domestic and international regulatory regimes, to some extent, have been responses to mostly corporate interests seeking to take advantage of emerging technologies. Again, this forces-relations relationship is dialectical. New technologies do not determine new social relationships. Technologies and techniques always are developed and applied by human beings acting on the basis of intersubjective conceptualizations of reality and these are shaped by various relations of production and a complex of mediators. More to the point, structured conditions and/or various acts of resistance condition the ways in which new technologies and techniques are adopted or rejected.[2]

Consumption ideals and practices in any given place and time play an important role in the complex structuring of human relationships. Moreover, the institution of consumption itself is directly influenced by other institutions as well as various organizations and technologies. The law, for example, particularly property law, is an institution that shapes consumption. It provides individuals with state-sanctioned rights in relation to commodities while rules, regulations and social norms, to varying degrees, shape how consumption takes place (Cosgel 1997, 155). Organizations, such as corporations, nation states, unions, NGOs, and many others influence ongoing consumption developments through a vast range of activities. Technologies (including techniques), such as the printing press, telesatellites, and even tools as banal as eating utensils, also influence the ongoing history of this institution. In sum, consumption is not only an institutional construct shaped by organizations, technologies and other institutions, it is itself an important component in the ongoing and complex construction of reality.

THE DEEPENING AND WIDENING OF CONSUMPTION

The task of modifying the institution of consumption (the task of, in effect, producing consumers) for the most part is being pursued directly by corporations and indirectly by states. In effect, corporations and states have mobilized various organizations and technologies—as well as other institutions (such as the law and religion)—in the task of modifying and expanding consumption. In relatively advanced economies, this has involved new efforts, through technology, to deepen already entrenched ways of thinking and acting. This also can be termed the commoditization of daily life in which "more aspects of family life, religious practice, leisure pursuits, and aspects of nature" are becoming commercialized and brought into the fold of capitalist relations (Gill 1995, 409). As for relatively less developed areas of the world, efforts to widen capitalist consumption activities into previously under-exploited and geographically unreachable markets continue.

This deepening and widening of consumption is a response to the dynamism of capitalism writ large and to recent and unprecedented developments in information and communications capabilities and investments. As Marx demonstrated, capitalists generally are so productive that they are forever facing the prospect of not securing markets for their goods and services. At its worst, "there breaks out an epidemic that, in all earlier epochs, would have seemed an absurdity—the epidemic of overproduction" (Marx and Engels 1979, 86). Competition over limited markets has involved efforts to produce and distribute commodities more efficiently through technological improvements. Recent manifestations of these developments constitute yet another stage in the systemic drive to expand consumption activities through, 1) institutional widening, 2) the "production of new needs" and the "discovery and creation of new use-values" through consumption's deepening, and 3) the "creation of new needs by propagating existing ones in a wide circle" through both widening and deepening (Marx 1973, 408).

In response to the rise of global (especially Asian-based) competition, the relatively high costs of unionized labor in the West, and other factors, in the 1970s mostly U.S.-based corporations came to recognize that production process efficiencies could be improved significantly through the development and large-scale application of computer-based technologies. Such technologies involve costs that eventually must be paid for through expanding profits. New technologies also may be applied in the development of new commodities that themselves require markets in which they can be sold. Finally, new technologies implemented in order to expand markets through, among other things, an extension of geographic reach, more effective marketing practices, and faster turnover times, also involve overhead costs that must be paid down and these include the often incalculable costs of reorganizing a range of production, distribution and exchange-related activities.

In sum, the systemic pressure of capitalism to be technologically innovative ultimately involves the need to expand profits and, more particularly, the economic compulsion to implement such innovations in an effort to substantiate or pay down their costs (Lachmann 1956). This, in turn, implies some form of growth in rates of consumption and this involves deliberate efforts to expand markets through the intensification of consumption activities (what I am calling its *deepening*) and/or efforts to grow markets by selling commodities in new territories or spaces (what I call consumption's *widening*).

The past twenty years of marketplace deregulation in the United States, followed by American-led efforts to open up and liberalize foreign markets through international trade agreements and related interventions, reflect extraordinary productive forces being developed in response to systemic pressures. While technological and other innovations were initiated by mostly corporate interests, of necessity, states—the sovereigns of legal authority in international relations—have been the essential mediators of efforts to restructure the way in which global capitalism (including consumption) takes place. Enormous investments and long-term corporate and state-based strategic plans today are directly linked to the need to construct a world in which new technological capabilities can be fully exploited and this involves rapid, pervasive, and problematic efforts to deepen and widen the institution of consumption.

The ideal conditions in which consumption may be widened involve both practical and ideological components. These include a secure and steady growth in production, wages, and the availability of commodities. It also involves a mass recognition that consumption and consumerist values are inherently good. Technological developments, in the workplace, in the marketplace, and in the home, play an essential role in all of these. Consumption as a facilitator for ongoing economic expansion also involves spatial transformations. Examples of this include the changing organization of homes and transportation networks to accommodate emerging consumption norms. In most Western industrialized countries (but in North America especially), suburbanization, the ascendancy of the shopping mall, and the massification of the automobile, along with the predominance of commercial television in the home, facilitated a dramatic growth of consumption among a burgeoning post-war middle class. From 1975 to 1998, real consumption in the world doubled to $24 trillion while 86 percent of all private consumption is carried out by just 20 percent of the world's population (UNDP 1998, 1–2).

This growth of consumption in mostly First World countries continues to involve a historical process in which diverse and complementary nodal points—organizations, technologies, and other institutions—facilitate (and sometimes retard) the development of consumerist thoughts and practices.[3] In addition, complementary nodes support the diversification and specialization of production, distribution, exchange, and consumption. Such developments, promoting the flexibility and even the individualization of consumption, ideally involve technologies that provide for a faster turnover of commodities directly as a result of falling spatial and temporal barriers separating producers from consumers.

This deepening and widening process, and the struggle to minimize spatial and temporal barriers, has involved technology-related developments such as the promotion of consumer credit capabilities, the growth of commercial mass media, the expanding reach of advertising and marketing interests and, of course, a range of state-based reforms aimed at regulatory, educational, and infrastructural developments (both domestically and internationally) crafted to facilitate rising rates of consumption (Lee 1993, 84). Keynesian macroeconomic policies, for example, still aspire to maintain stable growth rates in domestic and world consumption while, more recently, policies have been implemented to aggressively commercialize the Internet. The goal of this latter development is to ensure that the communications revolution is a boon to corporate interests and, more generally, the Internet becomes fully exploited as a new productive force for global capitalism (Baran 1998, 125–7).

Technological developments and applications, and, hence, large-scale investments, are essential in efforts to widen and deepen consumption. In relation to its deepening through the use of information and communications technologies, commodities such as financial services and entertainment products, and a range of specialized technologies such as personal computers, involve the promotion of a somewhat different ideological message than that related to widening. Rather than widening's general massification of consumption and its "keeping up with the Joneses" ideal, deepening more directly involves the acquisition and use of commodities as a means of enhancing the liberal ideal of individualized identity and meaning.

In seeking new markets, corporations attempt to become a growing presence in comparatively underdeveloped parts of the world. In already relatively saturated areas, many are compelled to go further and become an integral part of people's lives. Worldwide networks of instantaneous and interactive flows of information about consumer spending patterns and preferences are essential for this to proceed. Through digital communications, coupled with increasingly sophisticated database information on credit card and other spending activities, commercial interests monitor individuals. More participation online now means increasingly accurate information for vendors who, in turn, can more directly target prospective customers. The use of software on Internet browser systems that tells websites who is visiting (a technology called the *cookie*) is being used as a means of developing consumer profiles.[4] As a result, specifically targeted advertisements and icon come-ons are posted on the screen of the individual virtually every time he/she logs on and often throughout his/her visit to cyberspace.[5] The deepening of consumption also involves the rise of nonmaterial commodities characterized by unprecedented flexibilities, a plurality of consumer applications, and fast (if not instantaneous) turnover times. With the Internet and related developments, the institution of consumption now has the technological capacity to become, more than ever before, a deepening presence in our day-to-day lives.

CONSUMPTION THROUGH THE INTERNET

Arguably, no assemblage of contemporary productive forces better illustrates the compulsion to deepen and widen the institution of consumption than does the Internet. The advantages of its development and implementation for the ongoing development of capitalism are manifold. So too, however, are its potential contradictions.

The Internet is becoming a mass *and* specialized interactive marketplace. Its infrastructure is a virtually seamless web of information and communications technologies and services whose physical reach is limited only by the availability of the electricity and hardware needed to transmit and process its digital signals. Its presence enables corporations and entrepreneurs to sell or lease non-material commodities to anyone almost immediately and regardless of location. The capacity to integrate such services and tailor them to the particular demands of consumers, as well as to gather information for future marketing efforts, is unprecedented. For the producers of both material and nonmaterial commodities, the Internet furthermore constitutes an opportunity to eliminate bricks-and-mortar retailers.

Just as important as the ability to reach new markets through a virtual rather than a physical presence is the opportunity provided by the Internet to speed up turnover times toward the goal of what Marx famously referred to as "the twinkling of an eye." In addition to these incentives to make use of Internet technologies for commercial purposes is the necessity of paying down the costs involved in their implementation. For example, the economic (and other) costs related to the construction of the Internet already appear to be compelling an extraordinary corporate and government-led promotion of its economic and even democratic potentials. This has included an initiative, led by the U.S. executive branch, to make the Internet a tariff-free zone for transnational commerce (The White House 1997 and U.S. Department of Commerce 1999).

Investments made over the past decade or two in public telecommunications infrastructures are mounting. These are the wires and satellite links through which Internet commerce takes place.[6] Ongoing merger and acquisition activities among information, communications, and other corporate interests almost all involve strategic considerations related to prospective Internet developments. Such activities generally reflect the presence of a widespread demand for ever-improving information and communications capabilities primarily for the economic reasons outlined above. These investments do not, however, reflect significant shifts in the demands of consumers in either developed or relatively less developed parts of the world (at least not yet). While, historically, the relationship between corporate demand and the supply of information and communications facilities has been largely dialectical, with the Internet, for the most part, individual consumers have been relatively inactive, following supply far more than demanding it.

Examples of this effort to modify demand include the emerging practice among banks in relatively affluent communities to charge clients service fees if they want to deal with flesh-and-blood tellers. This is being pursued as a means of reducing labor costs and to impel otherwise reluctant customers to participate in electronic banking. The introduction of digitalized television services is another instance of corporate efforts to modify cultural norms. In the United States, an extraordinary consensus has been reached among a broad range of corporations on the technical standard on which to digitalize domestic television signals by the year 2006. In effect, U.S. corporate interests, through the legal and policing powers of the American state, hope to bring the Internet into people's homes through their television screens. In the United Kingdom, digitalized television broadcasting has been established and its main proponents—led by SKY Digital (controlled by Rupert Murdoch's News Corporation International)—consider it to be the prelude to a universally accessible and very commercial TV-based Internet system. More generally, efforts to get people culturally "wired" in relation to the Internet—especially the young (through the use of Internet-linked computers in schools) and those consumers possessing healthy disposable incomes—are well underway (U.S. Department of Commerce 1998).

One of the more remarkable features of plans to capitalize on Internet technologies involves the ability to, in effect, eliminate turnover time through what has been referred to as the virtualization of capitalism. Some software products, for example, constitute potentially ideal commodities for capitalists. Not only can such nonmaterial commodities be distributed and consumed almost instantaneously, unlike automobiles or screwdrivers, the makers of software products like Doom or Eudora can add layers and upgrades with little risk of saturating markets. One common strategy, as pointed out by Ronald Diebert, is to "release 'beta' and 'shareware' versions for free over the Internet to entice consumers into the latest versions (and also download the costs onto the consumer of product testing!)" (1998).

In the words of Veblen, "The highest achievement of business is the nearest approach to getting something for nothing" (1964, 92). Of course, ideals rarely translate into realities and even consumption over the Internet involves inevitable bottlenecks. After all, the more ephemeral brand of capitalism that technology hath wrought is still capitalism and, as such, tensions and outright contradictions are almost as predictable as the systemic compulsions underlying them.

TENSION AND CONTRADICTION ON THE INTERNET[7]

In the United States, the epicenter of global capitalism, from 1980 to 1992, private final consumption expenditures on recreational, entertainment, education, and cultural services increased (in constant prices) at an annual rate of almost 7 percent while total private consumption expenditures rose by 4.3 percent (United Nations 1993, 2007). A relatively small portion of the population has fuelled these growth rates. For most, real wages continue to stagnate or decline while wage and job security disparities between rich and poor grow (UNDP 1998, 29–30). Personal credit and working longer hours have enabled others to participate in the consumption boom but for many this has resulted in mounting debt. Perhaps not surprisingly, in 1996, 55 percent of Internet users in the United States were from the wealthiest one-third of the population (Media Dynamics 1997, 168). For reasons involving spending capabilities alone, the rapid formation of a mass consumer market for the products and services now being readied for the Internet is by no means a certainty.

As for the time to consume online, the average American has more free time today than ever before.[8] From 1965 to 1985, average annual gains in the free time available to full-time employed men in the United States was 8 minutes each day while for women it was 19 minutes. Similar gains in free time have been recorded in other industrialized countries (Robinson 1991, 138, 140).[9] More interesting than these averages is how people have been using this time. Through these decades, television watching has taken up increasing amounts of free time. In 1970, the television set was on for an average of 32.5 hours each week in U.S. households. In 1980, this figure rose to 46.5 hours. In 1995 the television was on for 50 hours a week (Media Dynamics 1997, 26). Not only has television viewing outpaced the growth of free time, time spent watching TV has increased more than any other free time activity such as reading, listening to music, or visiting friends. Surveys on the use of home computers indicate that the time an individual spends online generally reduces time spent using other electronic media, including television (Robinson and Godbey 1997, 165). For example, in the first two months of 1996, watching television was the most adversely affected free time activity among five hundred U.S. adults directly as a result of their use of personal computers (Media Dynamics 1997, 178).

Although the ability to use several forms of media at one time (for example, listening to a radio while "surfing" the Internet) reflects the complexity of time-use analyses, a cursory reading of such studies indicates that the free time needed to directly participate in Internet-mediated consumption is far from unlimited and may well be cannibalistic in relation to other electronic commerce activities. Proponents of the Internet counter such observations by pointing to the flexibility, efficiency and overall speed in which Internet commerce takes place as evidence that more can, and will, be done in less time. Gross time measurements, some argue, are secondary to what can be achieved over a given period of time. For example, more free time (theoretically) will be gained as a result of shopping online instead of driving to and walking through a shopping mall to make a purchase. But this line of reasoning assumes that consumers are "rational" actors who will both embrace timesaving technologies and enthusiastically participate in Internet commercial transactions.[10]

Television watching and Internet surfing are becoming remarkably similar activities. Indeed, as already mentioned, these may well merge into one medium. This should not be a surprise. The core source of revenue for both, at least at present, is their ability to attract people to advertisers. Like television, online sensations (involving seductive point-and-click icons, for example) arguably have become more important motivators for Internet use than some kind of purposeful quest for information. To attract eyeballs to advertisers and, hopefully, commercial websites, the Internet is becoming a kind of interactive million-channel digital television system specializing in endless provocations.

In addition and in relation to this effort to deepen consumption through the Internet and related technology-based developments is the emergence of an ever-changing and, arguably, more ephemeral culture, locally, nationally and even globally. This cultural shift perhaps has contributed to the common assumption that people have less, not more, free time. More time, when accompanied by more choice and immediacy, does not necessarily compel the development of a world characterized by ever-growing consumer desires. The speeding-up of life and its psychological implications instead could stimulate (and, arguably, already is stimulating) countercultures in search of less consumption, flesh-and-blood instead of online relationships, and perhaps even a greater valuation of the spiritual and community aspects of life over the commercial and individualistic.

Beyond such questions concerning income and time, the success of Internet developments thus also involves the capacity of cultures and individuals to undergo rapid modifications in their consumptive practices. Given the entrenched nature of consumption as a sociological institution—itself sustained through a plethora of related institutions, organizations, and technologies—the speed with which relatively well-off consumers are expected to modify their day-to-day norms in relation to investments involving Internet and related technological applications deserves some attention. In this context, the Internet may constitute a consumption crisis in the making. Questions concerning the ability of consumers to pay for new services, the time needed to participate, and their psychological-cultural willingness to live more and more of their lives in cyberspace remain unanswered.

In relation to the systemic compulsion to commoditize virtually every aspect of life, the institution of consumption has long been a significant component in constructing the meaningfulness of life. Consumption can, and usually does, link us to others in relationships that hopefully make sense. The role of consumerism as a meaningful ideology and consumption as a meaningful activity has become particularly significant given the cultural conditions of early twenty-first century capitalism. Consumption, for many, now involves much more than the satisfaction of subsistence needs. As technology enables capitalism to produce more, for relatively wealthy consumers whose consumption activities involve the search for meaning and identity, one can argue "that the struggle to maintain or expand profits is increasingly the struggle to control not just consumption practices but also the meaningfulness of the consumption experience itself" (Lee 1993). One of the cultural characteristics of contemporary capitalism is the emergence of a disparate but nevertheless global consumerist society "in which ever-growing consumption becomes the principle aspiration, source of identity, and leisure activity for more and more of the population" (Ackerman 1997, 109).

In this context, corporate interests will intensify their efforts to sell not just commodities but, more essentially, meanings. There are contradictory implications related to this. The systemic compulsion to control meaning (that is, the need to sell more "sizzle" in order to sell more "steak") will involve corporations in ever-larger and more complex mergers and partnerships involving digital technologies. The goal for such interests is to become a widening and deepening presence in workplace, household, transportation, recreational and other environments so that, in effect, the institution of consumption mediates virtually every moment of our lives. But as a result of the flexibilities garnered through new technologies and efforts to deepen consumption activities, consumption itself is becoming increasingly personalized. In effect, affluent consumers around the world are being encouraged to tailor their demands and consumption activities to suit their own preferences. However, such efforts to deepen consumption through new, flexible, and increasingly personalized technologies paradoxically may be making this systemic compulsion to control consumption a more, rather than less, difficult task. With increasingly sophisticated computer and digital technologies, resourceful consumers are becoming more capable of, in effect, manufacturing their own software and even hardware commodities thus potentially enabling them to circumscribe corporate efforts to control the meaningfulness of consumption experiences.

While a range of institutions, organizations, and technologies propagate and mediate the general ascendancy of consumption and consumerism, it is a mistake to underestimate the fact that some institutions and organizations, in effect, oppose the idealization of a consumerist society. Religious organizations (some becoming increasingly globalized through migratory developments, television transmissions, and even celebrity promotions) sometimes directly challenge aspects of the institution of consumption. Another institution—marriage—continues, for the most part, to promulgate ideals such as love and sharing over selfishness and materi-alism. But against these (and others), the mediators promoting an increasingly global consumerist society—and thus the institution of consumption—for the most part are inter-linked, mutually supportive, and generally overwhelming. De-spite this, given the limitations and contradictions characterizing consumption, it is not surprising that deviant behavior and pockets of organized opposition are ever present.[11]

CONCLUSIONS

In this chapter I have addressed the role of technology in shaping the institution of consumption. Not only does consumption remain an under-assessed aspect of the world economic system,[12] it constitutes an essential component of the ongoing development of global capitalism. Emerging alongside investments in new tech-nologies are vested interests seeking the reconstruction of consumption ideals and practices. In the world's peripheries these mostly involve efforts to widen con-sumption and in relatively developed social-economic contexts the emphasis is on its deepening.

Consumption can be conceptualized as a sociological institution—a histori-cally constructed and power-laden typification of habitualized thoughts and ac-tions—that directly shapes the thoughts and activities (or inactivities) of the innumerable agents of global governance. (On related concerns, see the chapter by Braman.) Its complex components are interlinked and sometimes challenged by other institutional constructs as well as various organizations and technologies. Consumption also constitutes a moment in the ever-dynamic capitalist production process. As such, the institution of consumption is the subject of ongoing, neces-sary, but often problematic modifications. Emerging productive forces—generally new ways of "doing" capitalism (including new technological applications)—compel this institution to be paradoxically both stable and dynamic. Established modes of consumption, and other institutional, organizational, and technological mediators that contribute to how reality is conceptualized and practised, usually place con-straints on the capacity to modify consumption. In the contemporary global polit-ical economy, remarkable demands are being placed on this institution and, as discussed, the Internet is a technology-based illustration of potentially contradic-tory developments involving its deepening and widening.

While Veblen addressed the relationship between technologies and insti-tutions at the end of the nineteenth century, Marx and Engels published one of the first analyses of technology and globalization—*The Communist Manifesto*—fifty years before that. "The bourgeoisie," said Marx, "cannot exist without con-stantly revolutionizing the instruments of production, and thereby relations of production, and with them the whole relations of society." Marx went on to write that "The need of a constantly expanding market for its products chases the bourgeoisie over the whole surface of the globe. It must nestle everywhere, settle everywhere, establish connections everywhere" (Marx and Engels 1979, 83). Herein I have argued that the institution of consumption is an essential but not well understood aspect of developments related to the Internet, governance, and, more profoundly, the future of capitalism and world order. Consumption, fundamentally, is a complex nodal point in the ongoing construction of reality and it is at this level of human consciousness that the increasingly technology-dependent edifice of global capitalism is both demonstrably dynamic and potentially vulnerable.

NOTES

Thank you to the editors and referees of this volume and my fellow contributors for their critical comments and helpful suggestions. I am especially grateful to Ronald Deibert and Jeffrey Hart.

1. Veblen elaborates as follows: "habits of thought are the outcome of habits of life . . . the discipline of daily life acts to alter or reinforce the . . . received institutions under which men live. And the direction in which, on the whole, the alteration proceeds is conditioned by the trend of the discipline of daily life" (1919:314).

2. Although this summary of the forces-relations of production dialectic is a much-simplified version of that developed by Marx, it is fair to point out an important shortcoming. In recognizing that the general level of development of the productive forces in a society explains why certain relations of production, and not others, can advance—and that these relations of production then facilitate the advancement of particular productive forces while retarding the development of others—precisely how this process works, for the most part, remains unanswered. As such, Marx's forces-relations dialectic provides a general explanation (a starting point, perhaps) but lacks precision in terms of the mechanisms at play at any particular time and place. On the subject of technology, it is necessary to complicate matters by recognizing that technology, in practice, almost always constitutes both a productive force and relation. A technique, for example, constitutes the essential ingredient shaping technological applications and to some extent it is also an outcome of such applications. A technology and the technique used in its application are virtually inseparable—one would be irrelevant without the other. It is in this sense that technology simultaneously is a productive force and a relation. This is not to say that in a general historical context a productive force should not be given analytical primacy: both empirically and logically, a productive force or at least the potential for its development must exist in order so that social and economic relations have at least the capacity to develop. On the logic of this, the following historical questions posed by Marx for his own ruminations are illuminating: "is Achilles possible when powder and shot

have been invented? And is the Iliad possible at all when the printing press and even printing machines exist? Is it not inevitable that with the emergence of the press, the singing and the telling and the muse cease, that is the conditions necessary for epic poetry disappear?" (1984:150).

3. An example of this complex relationship among institutions, organizations, and technologies in the history of consumption is the development of the modern commercial Christmas—a religious holiday that has become a significant component of the institution of consumption. A related development is the department store. Dating from the latter half of the nineteenth century, for the first time one retailer sought to sell customers everything. To encourage people to modify their established consumption habits, department stores were designed to promote a comfortable and positive shopping experience, particularly for middle class women. Department stores, in the decades preceding the automobile, both reflected and contributed to a middle class metropolitan lifestyle. Christmas norms were directly modified by the promotion of gift giving as mediated through the modern department store—a custom largely unknown before the middle of the nineteenth century. Subsequent developments, transforming Christmas into an annual orgy of mass consumption (Nissenbaum 1997), underlines the complex relationships at play involving consumerism, religion, family, mass media and other nodal points of social-economic reality (Ackerman 1997:116).

4. More precisely, the cookie is a text file saved in Internet browser directories or folders and stored in an individual's RAM while his/her browser is running. Most of the information in a cookie is technical and innocuous but websites also use cookies to record personal preferences.

5. Recently, for example, within two weeks of opening a new Internet account and after two visits to the Internet bookstore Amazon.com, I was greeted with a prominently displayed icon message reading "Hello Edward Comor" followed by an invitation to visit Amazon again and again.

6. In OECD countries, from 1980 to 1995, these investments have increased from $95 to $118 per capita (using 1995 prices and exchange rates). In the US, this figure rose from $111 to $124. In total (again using 1995 prices), in the United States these investments stood at $117 billion while in 1995 they totaled $191 billion (OECD 1997).

7. An elaboration of parts of this section can be found in Comor 2000.

8. Free time constitutes all time spent in activities (or inactivities) that are not obligatory. Obligatory categories of time use include work time, committed time (egs. family care, shopping, housework) and personal care (egs. sleeping, eating, grooming). The reader is reminded that data representing the average does not represent important disparities in the distribution of income, credit and time capabilities.

9. There are two main reasons for this increase in free time. First, fewer people are married and this has reduced time spent on family care and housework. Second, most workers in these relatively affluent societies are retiring earlier or are working fewer hours as they get older (Robinson and Godbey 1997: xvi).

10. The reader will note the use of quotation marks for the word rational. The purpose is to underline the inter-subjective nature of rational (including consumer) thought and action.

11. The individual raised in relative affluence, for example, recognizing that the cliché "money can't buy you happiness" resonates with his/her experiences, may reject truths equating consumption with a meaningful life. This way of thinking and subsequent practices constitute a direct challenge to the norms of contemporary consumption. Because of the range of mediators challenged by such anti-consumerist attitudes and activities, a range of institutions, organizations, and technologies will directly or passively rebuke or isolate these malcontents. However, barring revolution, the only mediators that have successfully challenged consumption norms on a large scale and over extended periods of time have been compelled to become participants in consumer society. Organic farming for instance—once (and still) promoted as an alternative to mass consumer factory farming—has become big business in its own right involving the marketing and sales of how to books and specialized health food shops. Even environmental organizations are compelled to market the natural environment to prospective donors through glossy promotions focusing attention on the plight of baby seals and other attractive, irresistible, and meaningful victims. Predictably and paradoxically, even Greenpeace accepts donations charged to credit cards (Luke 1998).

12. As John Kenneth Galbraith recently has pointed out, "On no matter is economics more in contradiction with itself than in its view of consumer behavior and motivation and the consumer-oriented society" (Galbraith 1997: xxi).

REFERENCES

Ackerman, Frank. 1997. The History of Consumer Society. In Neva R. Goodwin, Frank Ackerman and David Kirton, eds. *The Consumer Society*. Washington, D.C.: Island Press.

Baran, Nicholas. 1998. The Privatization of Telecommunications. In Robert W. McChesney, Ellen Meiksens Wood and John Bellamy Foster, eds. *Capitalism and the Information Age*. New York: Monthly Review Press.

Berger, Peter L., and Thomas Luckmann. 1967. *The Social Construction of Reality*. Garden City, N.Y.: Anchor.

Cohen, G.A. 1988. *History, Labour and Freedom*. Oxford: Clarendon.

Comor, Edward. 2000. Household Consumption on the Internet. *Journal of Economic Issues*. 34.

Cosgel, Metin M. 1997. Consumption Institutions. *Review of Social Economy*. 55.

Cox, Robert W. 1995. Debt, Time, and Capitalism. *Studies in Political Economy*. 48.

Deibert, Ronald J. 1998. Private correspondence with author.

Gill, Stephen. 1995. Globalization, Market Civilization, and Disciplinary Neoliberalism. In *Millennium*. 24.

Galbraith, John Kenneth. 1997. Foreword. To Neva R. Goodwin, Frank Ackerman and David Kirton, eds. *The Consumer Society*. Washington, D.C.: Island Press.

Lachmann, Ludwig. 1956. *Capital and its Structure*. London: G. Bell and Sons.

Lee, Martyn J. 1993. *Consumer Culture Reborn*. London: Routledge.

Luke, Timothy W. 1998. The (Un)Wise (Ab)use of Nature: Environmentalism as Globalized Consumerism. In *Alternatives*. 23.

Marx, Karl. 1973. *Grundrisse*. Harmondsworth: Penguin.

———. 1984. Introduction to a Critique of Political Economy. In Karl Marx and Frederick Engels. *The German Ideology*. New York: International Publishers.

Marx, Karl, and Friedrich Engels. 1979. *The Communist Manifesto*. Harmondsworth: Penguin.

Media Dynamics. 1997. *TV Dimensions '97*. New York: Media Dynamics.

Nissenbaum, Stephen. 1997. *The Battle for Christmas*. New York: Vintage.

OECD. 1997. *Communications Outlook 1997*. Paris: OECD.

Robinson, John P. 1991. Trends in Free Time. In Wendy O'Conghaile and Eberhard Kohler, eds. *The Changing Use of Time*. Dublin: European Foundation for the Improvement of Living and Working Conditions.

Robinson, John P., and Geoffrey Godbey. 1997. *Time For Life*. University Park: Pennsylvania University Press.

UNDP. 1998. *Human Development Report 1998*. New York: Oxford University Press.

United Nations. 1993. *United Nations Statistical Yearbook*. Issue 40. New York: United Nations.

US Department of Commerce. 1998. *The Emerging Digital Economy*. Washington, D.C.: US Government.

———. 1999. *The Emerging Digital Economy II*. Washington, D.C.: US Government.

Veblen, Thorstein. 1919. *The Place of Science in Modern Civilization and Other Essays*. New York: Viking.

———. 1953. *The Theory of the Leisure Class*. New York: Mentor.

———. 1964. *The Vested Interests and the Common Man*. New York: Sentry Press.

The White House. 1997. *Framework For Global Electronic Commerce*. Washington, D.C.: US Government.

PART III

GOVERNANCE IN TELECOMMUNICATIONS

CHAPTER EIGHT

CAPITALISM, TECHNOLOGY, AND LIBERALIZATION: THE INTERNATIONAL TELECOMMUNICATIONS REGIME, 1865-1998

MARK W. ZACHER

INTRODUCTION

Open and coordinated communications flows between countries are vital for international firms, governments, and national populations. This chapter outlines the promotion of commercial openness from the mid-nineteenth century to the present. It focuses on regulations concerning jurisdictional rights, technical standards, as well as market access and pricing, and it explores the influence of three general factors that have been the central driving forces of economic liberalization.

The politics of international telecommunications relations and particularly of international cooperation are fascinating for a variety of reasons. First, in 1865 the International Telegraph Union (the predecessor of the International Telecommunication Union or ITU) was the first global international organization to be created. Second, the evolution of international telecommunications provides an important story of growing economic liberalization over the last century and a half. That is to say, there has been progress toward the reduction of barriers to the flow of international communications. There have always been some important forms of international cooperation in jurisdictional and technical areas to facilitate the flow of communications, and the scope of this cooperation has grown markedly over the last half of the twentieth century. With regard to the issues of market access and pricing, international cooperation actually hindered liberalization from the mid-nineteenth century through to the late twentieth century, but these old forms of cooperation, which were based on a cartel among

The author would like to thank Hilla Aharon, Jim Rosenau, J.P. Singh, and Debora Spar for their comments on previous drafts.

state telecommunications administrations, began to disintegrate in the 1980s. Through means such as divestiture, deregulation, re-regulation, and privatization, states have transformed their telecommunications services industries from tightly controlled monopolies into competitive markets. A discussion of alternative liberalization policies falls out of the realm of this discussion. However, it should be noted that privatization does not necessarily imply liberalization, and conversely liberalization may include one or more government-owned entities (Hills 1986). Although currently, states have very diverse international telecommunications regulations, there were signs in the late 1990s that the international community might be in the process of refashioning a regime for market access and pricing based on a high degree of economic liberalization.

Third, the liberalization components of the international regime have been shaped by three general factors: the capitalist world political economy, the distribution of state power (particularly the existence of a dominant state or group of states), and technological change. Capitalism encourages firms to seek business opportunities throughout the world, and this drive on the part of firms has been fundamental to intergovernmental efforts to maintain and increase commercial openness for their firms. In a sense, large firms throughout the world have operated under the mandate of interconnection in that they have consistently demanded technical interconnection among commercial enterprises throughout the world. At the same time, it is questionable whether changes in capitalism have been mainly responsible for changes in the international telecommunications regime. For example, the drive toward deregulation in capitalist states since the 1970s has facilitated liberalization, but it is doubtful whether it was largely responsible for the significant changes in international telecommunications politics in recent decades. It is unquestionably the case that the United States expedited the liberalization process—perhaps by a decade or more. It is, however, doubtful whether American interests and influence are the central forces behind the long-term movement toward greater commercial openness. What has been fundamental to the small and large changes in the international telecommunications regime has been the evolution in international telecommunications technology. Technological innovations have led states to adopt frequent small changes in matters such as technical interconnection standards and the allotment of the frequency spectrum to different services, and they have also been responsible for the dramatic alterations in the politics and law of international telecommunications in the last several decades of the twentieth century.

The purposes of this chapter are to trace the trends toward liberalization or economic openness in the international telecommunications regime and to analyze the major factors that have shaped the regulatory trends. The first part of this chapter focuses on progress with regard to jurisdictional rights and technical standards, while the second part looks at the sphere of greater change—market access and pricing. In the course of this analysis of regime change, particular attention is given to the impacts of evolving technology, changes in capitalist ideology and practice, and the distribution of state power (particularly the existence of a hegemonic state). All of them have had an influence, but technology has been the central shaping force. From the international capitalist system there flows the mandate for interconnection, but the changes in international capitalism are not mainly responsible for the evolution of the telecommunications regime. The popularity of deregulation has had an impact on the telecommunications regime, but it has not been the major influence on regime change. Likewise, the United States has unquestionably played a very important role in promoting liberalization in recent years, but it has been firms' and governments' responses to the commercial benefits of new telecommunications technologies that have driven the evolution of the regime.

JURISDICTIONAL RIGHTS AND TECHNICAL STANDARDS

1865–1939

Telecommunications first emerged in the 1830s with the invention of the telegraph. It was, however, not until the 1850s that the telegraph was used extensively for communication between adjacent countries. Then, by the mid-1860s, the telegraph was employed extensively for transoceanic communications with the laying of a network of cables between the continents. During the 1850s and early 1860s, international business firms became more reliant on telegraph communications, and they objected to the host of technical barriers that hindered effective transmissions. In fact, the impact of telegraph communications on international business in the mid-nineteenth century was of a similar order as the impact of the merger of telecommunications and computing in the late twentieth century. Businesses thus began to press for international accords to create international standards that would eliminate many of the technical barriers. In 1865, most of the major western countries assembled in Paris and created the International Telegraph Union (ITU) with mandates to facilitate the flow of telegraph transmissions and to improve commercial coordination. The major areas of technical standardization concerned switches at *gateways* along international borders, international cables, and codes for communication. In subsequent decades ITU conferences concerned with technical standards and codes were factious because such accords involved adjustment costs and sometimes influenced states' comparative advantage in international markets. Nevertheless, agreements were concluded on the important obstacles that affected interconnection since effective interconnection was demanded by firms involved in international commerce. Clearly, the telegraph had marked impacts on the growth of trade and investment, and the growth of the international telegraph system was embraced by the commercial community (Codding 1972: 13–21, 65–67; Headrick 1991).

At about the turn of the century, international telecommunications greatly expanded with the introduction of two revolutionary technologies—telephone and radio. In the case of telephones the standardization problems were somewhat similar to those in telegraph communications as most of the standards that commercial interests demanded concerned technical guidelines for switches at international borders. Over time there was, however, an increasing need for standardization of both switches at borders and some types of equipment within countries. This led to the creation of international consultative committees on standardization with regard to telephone, telegraph and radio in the mid-1920s. The choices of technical standards for telephone switches, transmission equipment, and cables had some implications for the markets of national manufacturing firms, but overall the levels of standardization were rather modest in comparison to those required today (Codding 1972: 32–36; Chapuis 1976). It is important to note with regard to both telegraph and telephone systems during the late nineteenth and early twentieth century that the world was divided into three somewhat self-sufficient and technologically distinct networks: continental Europe, Britain and its empire, and the Western Hemisphere. The standards employed by the European countries were generally accepted for interregional communications, and also for some types of equipment used within the Americas and the British Empire. The international competition involved in standard setting at this time was muted by the fact that British and American manufacturers dominated the equipment markets in their regions. Also, in Europe there were several dominant national manufacturers that controlled particular segments of the equipment market.

The advent of international radio communications in the first decade of the twentieth century raised some regulatory problems that were similar to those that arose at the previous international telephone and telegraph conferences. It was necessary to standardize certain communication codes (as was the case with telegraph communications), and it was also necessary to standardize certain types of transmitting equipment in order to facilitate interconnection between radio stations or transmitters in different countries and on ships. One particular regulation is of interest because of the way in which the competition for the equipment market was handled. In the early years of the twentieth century, the Italian-British Marconi company tried to prevent ships and shore stations from using its equipment to interconnect with stations using equipment manufactured by other firms. In 1906, the newly created International Radiotelegraph Union (IRU) prescribed the norm that stations could not refuse to interconnect with other stations be-cause of the make of the equipment that those other stations were using. In other words, competition was not allowed to interfere with international interconnec-tion and hence maritime safety (Codding 1972: 83–96).

A unique problem raised by radio concerned the rules for the use of the fre-quency spectrum which, of course, falls within national as well as international airspace. States have never explicitly adopted an accord on the jurisdictional sta-tus of the frequency spectrum or airwaves within national airspace, although states have implicitly accepted that national stations have a right to transmit through foreign airspace—what one might call a right of innocent transmission through foreign airspace. There was, and still is, an implicit understanding that the frequency spectrum circling the planet constitutes a kind of common prop-erty resource. States would not have accepted a treaty with this explicit under-standing since they support national sovereignty over airspace for air travel. The jurisdictional understanding concerning a right of innocent transmission through foreign airspace was implied in the early twentieth century IRU rule (still accepted today) that states should not broadcast on any frequencies being used by foreign radio stations if they were registered with the IRU and were actively used by the foreign stations. Many transmissions, of course, use the airspace of several coun-tries (Codding 1972: 92–95; Leive 1970: 41–42).

During the early twentieth century, an IRU system for regulating the use of the airwaves was widely adopted. First of all, the IRU conferences *allocated* certain frequency bands to particular services such as maritime communications or pub-lic radio. Second, the conferences *allotted* specific bands to particular countries for purposes connected with the safety of international travel—that is to say, for mar-itime and aeronautical communications. There were also some national allotment plans that were accepted at a regional level to prevent interference among public radio stations. Third, and most importantly, the IRU *registered* particular frequen-cies for individual states if they could establish that they were not already in use. This system of IRU registration after national stations' successful use of them (first-come, first-served) was a bedrock of order and hence openness in interna-tional radio transmissions. An implicit part of the regulatory system was that na-tional stations would adjust their use of particular frequencies if advances in technology made it possible to use more frequencies in a particular band of the spectrum. It was also explicitly rejected that states could maintain rights to par-ticular frequencies after their national station ceased using them. In other words, efficient use of a common property resource and openness were placed ahead of states' attempts to extend their territorial jurisdiction. From 1865 through 1939, there were major technological changes and many new regulations, but interna-tional firms and states remained committed to interconnection (Codding 1972: 113–79; Savage 1989: 32–36, 67–71).

1945–PRESENT

The fundamental normative direction of the telecommunications regime with regard to jurisdictional and technical standards was established in the late nineteenth and early twentieth centuries, and it has remained quite stable in the last half of the twentieth century despite major technological changes. The central normative guideline has been the commitment to promote interconnection and hence the free flow of information. This norm has shaped the development of rules with regard to new means of communications such as satellites and fiber-optic cables as well as new telecommunications services such as facsimile and electronic mail.

The fundamental jurisdictional development in response to the invention of communications satellites was the acceptance of outer space as a common property of all nations. This was accepted in the Outer Space Treaty of 1967 which stated that outer space should "be free for exploration and use by all states" and should "not [be] subject to national appropriation by claim or sovereignty." There were certainly some countries that wanted to appropriate particular slots in the geostationary orbit (GSO); but the dominant consensus was that states should have free access to stationing satellites in the GSO (first-come, first-served). However, they also agreed that states should coordinate their activities to prevent intervention and provide access for all countries. Over time the satellite powers did develop coordination planning mechanisms for locating satellites, and they also accepted that all countries would have at least one slot. The combination of the jurisdictional status of outer space and various planning mechanisms has served the goals of openness and efficient use very well (Christol 1982: 3–58).

An interesting jurisdictional issue that consumed states in lengthy negotiations from the early 1970s through the early 1980s was whether states had to give their prior consent to the beaming of signals from direct broadcasting satellites (DBS) to their populations. Most of the developing countries pushed for the obligation to obtain prior consent, while the western states opposed it. In the end the developing countries generally gave up their political quest due to the fact that a satellite signal's range (*footprint*) covered so many countries. That is to say, consent to broadcast by one country led to receptions of the same satellite signal in dozens more. In addition, there was no way that states could jam satellite broadcasts in an efficient manner, and this encouraged many states to withdraw their early support for prior consent. Quite simply, international openness was promoted by the very nature of the technology (Christol 1982: 648–49, 702–09; Fjordbak 1990).

While some important developments occurred on the jurisdictional front in the last half of the twentieth century, it was in the realm of technical standards that states' concern for interconnection and the free flow of information had its greatest impact. The ITU Red Books of technical standards grew exponentially.

Perhaps the most important technological change was the integration of computers into telecommunications systems starting in the 1950s. This technological development required that states standardize many kinds of telecommunications equipment within their borders (e.g., cables and desktop computers) as well as at borders (e.g., switches). The integration of computer technology with telecommunications also led to two particular developments in the 1950s that required extensive international standard setting—the introduction of transoceanic telephone cables and the use of direct dialing. These developments were soon followed by the introduction of two new means of transmission (satellites and fiber-optic cables) and then subsequently a variety of new or greatly expanded telecommunications services. The international services include data transmission (an immensely important service for international business), fax, videotext, video conferencing, and email. These new services were built on technological standardization of many types of equipment throughout telecommunications networks. The ITU Secretary-General commented in the 1990s that "Standards making has clearly become the dominant collective activity in the telecommunications-information world today" (Tarjanne 1990). This comment is somewhat misleading in the sense that standard setting has probably been the most important activity of the ITU since the 1920s. The other major sphere of ITU telecommunications regulation has been control of market access and rates. Regulatory activities concerning this issue tended to take place more outside the ITU than within the ITU, and since the late 1980s, the ITU has only been marginally involved in this area.

The ITU occasionally failed in its attempts at standard setting over the last half century, but the key failures concerned consumer products such as television, high definition television, videotext and cell phones that are used for communication between services providers and consumers—not between commercial firms. States and their national champion firms could maintain different standards for such consumer products because their differences did not stand in the way of international commercial communications. When business communications have been at stake, users quite simply have demanded standards that permit the free flow of information (OECD 1988: 58–61). The international commercial system in a sense overrides states' concerns about maintaining national distinctiveness and policy autonomy in these spheres. Of course, states pursue standard setting for the simple reason that their firms and their entire national economies will suffer from exclusion from the global commercial information network if they do not accept interconnection standards. As several authors commented: "companies and countries can no longer afford to use a multitude of electronic machines in order to be able to talk to each other" (Bar and Borrus 1987: Sec. I). Another expert from France stated that "The restructuring currently underway has given the industry an international dimension that is incompatible with national standardization" (Bressand 1988: 293, Savage 1989: ch. 4).

There is extensive literature on factors that have influenced states' and firms' success on standard setting. These factors include the domination of a particular country or firm in an industry (Besen and Johnson 1986), the concentration of buyers in a single state (Cerni and Gray 1983: 48–49), economies of scale (Cerni 1984: 48), and an absence of a commitment to protect national manufacturers (McKnight 1987). These all have some relevance, but they exclude what is probably the most central factor—the relevance of technological standardization for international interconnection and competitiveness. The simple fact is that international business users of telecommunications will not tolerate impediments to interconnection. Of relevance is a comment in a study on technical standards: "In the nature of telecommunications technology there exists a strong imperative for technological homogeneity. Only in a very few instances can that imperative be overcome by domestic economic, social, and political concerns" (Savage 1989: 217).

While the best known area of technical standard setting in international telecommunications concerns equipment, there is another sphere of telecommunication standard setting that deserves noting—the management of the use of radio spectrum. What has happened in this sphere since 1945 is also basically consistent with the normative thrust of the regime that evolved between 1906 and 1939; but the post-1950 regulatory arrangements have expanded tremendously in complexity in response to technological developments. Technological improvements in transmitters permitted the use of higher and higher bands in the frequency spectrum, and this led to an increase in the number of services for which bands were allotted by ITU conferences. Also, the advent of satellite communications increased the number of telecommunication services, which meant that the ITU had to designate a larger number of services that were allotted bands. The conflicts over how wide a band and what band to allot to a particular service were, and still are, a matter of vigorous international conflict at ITU conferences, but in the end states agree on the allotments since failure to agree entails interference among transmissions and higher costs for commercial enterprises using the radio spectrum. Usually most states and interest groups are accommodated reasonably well in the spectrum plans so as to assure their support and to prevent interference. For example, in the 1970s, the ITU began to allocate individual frequencies within certain bands as well as particular geostationary orbital slots to developing states so as to assure their political backing. What the developing countries received was modest and was very much influenced by developed states' views of efficiency and their self-interests. Still the frequency spectrum management system is a remarkable political edifice that supports order and openness in international radio communications, and all countries back it despite some reservations (Codding and Rutkowski 1982; Savage 1989: ch. 3).

A recent development in telecommunications standard-setting is that an increasing number of standards are set by important national and regional bodies

(American, European and Japanese) or by private firms. For the most part the standards are eventually approved by the ITU, but sometimes they are not. A key reason for this development is that technology changes so rapidly now in telecommunications, it is difficult to obtain the consent of ITU bodies speedily enough for the telecommunications equipment market. ITU bodies have, in fact, been reformed to make their decision-making quicker, but it is still not fast enough in some circumstances. It is important, however, to recognize that the weakening of the ITU standard-setting bodies does not mean that there is an absence of standard-setting in international telecommunications. It rather means that the standards are set by a variety of powerful national, regional and global organizations that coordinate their activities through a number of channels. Technical interconnection and hence commercial openness are still assured by a complex network of government and industry officials (Cowhey 1995; Rutkowski 1995; Zacher with Sutton 1996: ch. 5).

An interesting dimension of technical standards that has arisen quite recently is the attribution of names and numbers for Internet users. Because the Internet was developed as a result of initiatives by the United States, the U.S. government and U.S. firms had significant impacts on its early regulatory structure. The Internet Corporation for Assigned Names and Numbers (ICANN) emerged in 1995 as the central international regulatory body, and it is a nongovernmental organization with the participation of private firms as well as governments and intergovernmental organizations. Despite its formal non-governmental character, states, and especially the United States, have had a strong influence on ICANN. The central concern of most commercial participants is the promotion of the global commercial transactions and the facilitation of the exchange of information, but this commercial concern might not be realized without the leverage of states. Governments do not allow the drive for profit by particular enterprises to disrupt international commercial exchanges. Governments, as well as most Internet firms, have been protagonists on behalf of the mandate of openness (Mueller and Thompson 2000).

In surveying the scope of jurisdictional understandings and, more importantly, technical standards for equipment and radio and cable transmissions, it is interesting to ask: what has driven the increasing output of regulations and their success in promoting interconnection and damage avoidance? The volume of regulations has to be attributed to the remarkable explosion of telecommunications technology which concerns new modes of communication, new services, and the integration of the telecommunications industry with other industries such as the computing, cable, and entertainment industries. Recent trends in deregulation and U.S. assertiveness did not significantly increase the volume of regulations although the United States did have an important influence on the utilization of and hence the regulation of outer space. American preferences were not significantly different from most industrialized states in ITU planning conferences. When one moves from explaining the volume of regulations to explaining the consistent normative thrust of international interconnection embedded in the international regime, one must look basically to international capitalism and the incentives it provides to firms to embark on international business ventures. If there is one thing that international firms desire, it is the ability to communicate with subsidiaries and customers throughout the world. States are, of course, responsive to the requests of their national firms and to their success in international markets, and they have been active supporters of rules of interconnection and damage prevention inside and outside the ITU. The state-business alliance has seldom been seriously strained in this area.

MARKET ACCESS AND PRICES

It is debatable whether the international arrangements for managing prices of telecommunications services and the market shares of service providers was liberal or protectionist prior to the 1980s, but before this issue is discussed, it is important to understand these arrangements that provided telecommunications services until very recently. In all states there was a telecommunications monopoly (generally state-owned) that controlled the sending and receiving of all transmissions. After the creation of the ITU in 1865, these state monopolies collaborated with regard to the establishment of market shares, rates, and certain commercial practices. The cartel system assured that each state monopoly collected fees for all international transmissions from that state. Also, the state monopolies were not allowed to compete for business with each other, and this was upheld by rules such as: (i) they could not alter the collection rates between two cities by altering the international route; (ii) they had to charge the same transit rates to all foreign telecommunications firms; (iii) and they could not interconnect with firms other than fellow state monopolies. In addition, they agreed to maintain joint ownership of all intermediary transmission technologies such as cables and satellites so that commercial challengers could be excluded from the industry.

The cartel system was defended by many governments and experts on the basis that a cartel of state monopolies was the most efficient system of providing international telecommunications services because there were economies of scale in national and international telecommunications systems. On the other hand, it was clear that state monopolies were charging rates for international transmissions that were significantly over cost so as to subsidize their domestic telecommunications systems. This fact certainly constituted a serious challenge to the economic efficiency argument. Also, it is of great importance to understand that, within the capitalist global economy in the late nineteenth and early twentieth centuries, there was a strong commitment by states to maintain state control of major infrastructure industries in communications and transportation. (An exception to this existed in the developing world where these countries could not afford their own telecommunications monopolies and had to rely on foreign firms such as Cable and Wireless and ITT to manage their telecommunications services.) States in this era saw both that the control of these infrastructure industries by the state was necessary for national economic development and that it also afforded state authorities control over intrusions by foreign states. Mercantilist thinking (including fears of economic vulnerability) had an important influence on the international capitalism of the day, and definitely impeded certain dimensions of commercial liberalization (Codding 1972: 14–60; Herring and Gross 1936; Headrick 1991).

The international cartel governing international communications remained very strong in the immediate post-World War II decades. In the case of the United States, there was not a state-owned telecommunications monopoly, but AT&T was designated to act as the monopoly firm controlling all international telephone communications into and out of the United States. An anomalous arrangement in the telecommunications regime was that several U.S. telegraph companies existed, but they were not allowed to disrupt the dominance of the intergovernmental cartel. In the late 1950s and 1960s, a number of U.S. telecommunications firms began to pressure the U.S. government to allow competition against the monopoly providers in particular regions and services. The government accepted limited forms of competition; however, this liberalization only concerned certain domestic services and did not intrude into international communications (Aronson and Cowhey 1988: chs. 1, 2, 4; Rutkowski 1995: 229).

Some modest, but important, changes in the international telecommunications industry started in the 1970s, and almost all of them originated in the United States. First, pressure developed from multinational corporations to lease lines from the monopoly providers so that they could obtain cheaper and better customized services. Firms recognized that the quality and price of their telecommunications services were crucial to their international competitive positions, and they became increasingly unhappy with the standard fare provided by the state monopolies. Large corporations, in particular, realized that making the most efficient use of telecommunications would result in substantial savings and productivity growth. The United States authorized leasing of lines by private firms, and this allowed large U.S. corporations to manage de facto their own telecommunications firms. This development of leased lines was a very important first step in firms' recognizing the advantages of competition in telecommunications. Other industrialized states slowly came under comparable pressure by their national firms, and they gradually succumbed to private pressure to allow leased lines

(Johnson 1986, 60–64). Second, ways were found by some international tele-communications firms such as AT&T to bypass the monopoly firms in foreign countries so they did not have to accept the rate structures of those foreign mo-nopolies. These bypass arrangements went under names such as *callback* and *third country calling*. Despite the attempts on the part of foreign states to prevent these bypass strategies, they had little success. Third, international consortia of firms from different countries began to develop since the member firms judged that they could best pursue their own interests by cooperating. These developments in the private marketplace basically sent out signals that international liberalization and competition were the wave of the future. These changes in the 1970s were cer-tainly due in part to the growing popularity of deregulation in professional eco-nomic, private business, and government circles as well as to the leadership of the United States. However, of central importance, was the evolution of technology, the emergence of new or improved services, and the increasing importance of telecommunications services to the competitive positions of firms (Zacher with Sutton 1996: 65–70).

While international firms secured concessions on market liberalization from the United States and some other industrialized countries, these states did not try to undermine the traditional state monopolies and the international cartel of these monopolies. This approach changed in 1980 with the United States's move to breakup AT&T into a number of regional service providers (Baby Bells) and an international service provider which retained the title of AT&T. Also, the market for international services was opened to competition. (MCI and US Sprint emerged in the US market before this.) The first countries to support the Ameri-can approach were the United Kingdom and Japan, and together these three coun-tries led the first wave of liberalization in the early 1980s. It was not until the late 1980s that the other developed countries also accepted varying degrees of liberal-ization, and for the most part the developing countries did not accept significant moves toward opening their markets to competition until the 1990s. Over the 1980s, the industrialized states recognized that if they did not provide firms within their borders with the same low-cost and sophisticated services that the lib-eralizing states were offering, foreign firms would locate in the liberalizing states (Horwitz 1989).

One of the remarkable developments of the 1970s and 1980s is that telecommunications users became much more active and influential than they had ever been. Price, efficiency, and quality of service were becoming increas-ingly more important to large corporations in information-intensive industries, the most notable being the banking sector (Wellenius et al. 1989: 79). Firms that were heavily reliant on telecommunications lobbied vigorously for less ex-pensive and more technologically refined services. They were particularly con-cerned with enhanced services such as data processing on which multinational firms depend a great deal, but they were also very interested in the liberalization of basic telephone service. Under the pressure of the lobbying and commercial practices of their firms, the developed countries began to violate the ITU rules that were at the heart of the traditional cartel so that, by the beginning of the 1990s, the old system looked very frayed. In fact, the national monopolies began to compete in each other's markets through a variety of means including callback and third country calling. Some attempts were made by states that favored the state monopolies and the intergovernmental cartel to enforce tight compliance with the ITU rules, but the major liberalizing countries and multinational business interests would not comply (Zacher with Sutton 1996: 167–72).

A very important step in the process of telecommunications liberalization was the conclusion of the General Agreement on Trade in Services (GATS) as a part of the Uruguay Round of the GATT (1986–94). Of considerable importance is that it required transparency with regard to states' competition and protection policies. Also, it established that foreign firms should have access to national telecommunications networks on reasonable and nondiscriminatory terms, and that equal national treatment be accorded to these firms. However, states' obligations to apply these norms were limited to those specific spheres of telecommunications where they made specific commitments, and in the 1994 GATS, these commitments were not very extensive with regard to telecommunications. Most states were quite simply not willing to grant open access to their basic telecommunications markets. On the other hand, states had agreed to general normative guidelines that pointed to an increasing acceptance of competition in their telecommunications markets (Sampson, 1996: 25–27; Nicolaides 1995; Fredebeul-Krein and Freytag 1997: 484–86). At the time of the signing of the agreements, and immediately thereafter, the developing countries did not support the provisions regarding telecommunications; but in the long run it is going to be difficult for them to reject the rules. A major pressure on the developing countries is that they desperately need telecommunications infrastructure and expertise for development purposes, and the most effective strategy for obtaining this infrastructure is through seeking investment by foreign telecommunications firms (Beltz 1996: 47). There are already marked movements toward liberalization in Latin America with decreases in state ownership of telecommunications firms and increases in competition among service providers, and comparable changes are occurring in Asia (Mody et al. 1995; Comor 1996; Ryan 1997; Singh 1999).

Because the GATS did not include a large number of specific commitments in telecommunications, the United States and some other industrialized countries pushed for a supplemental accord. A committee of the new World Trade Organization met on the subject from May 1994 through February 1997 when an agreement was finally signed. The accord, which is attached to the GATS, was approved by 54 states and the European Union. These states account for 90 percent of global telecommunications services. As the negotiations progressed over the three year period, an increasing number of states became supportive of liberalization in telecommunications because of pressure from their own firms and other states as well as a result of their own intellectual understandings of the advantages of liberalization. The United States exerted considerable influence in favor of competitive markets partly because the norm of open competition required that in order for foreign firms to enter the U.S. market their home governments had to provide equivalent competitive opportunities for U.S. firms (Beltz 1996, 45; Hufbauer and Wada 1997; Drake and Noam 1997). In the first two years of the negotiations, the European Union and the developing countries did not want to accept some of the U.S. proposals for liberalization, but by the third year they were much more willing to adopt U.S. proposals on the liberalization of enhanced services (e.g., data transmission) and basic services (e.g., telephone) as well.

Important regulations were accepted with regard to access for firms to foreign markets, openness of states to foreign investment in the telecommunications industry, pro-competition regulatory principles, and the implementation of transparency policies. Most developed states accepted extensive liberalization of all basic telecommunications services, and this, of course, involved allowing several suppliers to operate in their market. There are still some weaknesses in the agreement with regard to the specificity of regulatory principles, the criteria that states use in settling international accounts, and a number of other matters. In addition, state regulatory agencies are allowed to intervene in markets in order to assure universal service and the technical integrity of networks. With regard to the developing countries, concessions were made to allow them to phase in their liberalization commitments over approximately ten years. Despite these weaknesses or loopholes, significant progress toward liberalization was achieved. Central to the entire process was states' recognition that in order for their multinational firms to compete in the international economy, they must offer them low cost and sophisticated telecommunications services. In order to do this, these states must not only attract investment by the leading international telecommunications firms but must allow open competition among service providers (Drake and Noam 1997: 94–97; Fredebeul-Krein and Freytag 1997: 486–91; Singh in this volume).

While significant progress has occurred with respect to the liberalization of international telecommunications markets in World Trade Organization agreements as well as in states' legislation and regulations, the old state monopolies are still quite strong in many countries, and states continue to practice a variety of protectionist policies. In fact, some monopolies have been strengthened by entering into international alliances with each other, and in some countries like the United States the breakup of the old monopoly has been followed by mergers of several firms. While the old world of monopoly suppliers has persevered in varying degrees, major movements toward liberalization are under way in the developed world, and even in the newly industrializing countries, liberalization is moving ahead. The European Union has recently allowed access by cable companies, railways, and other utilities to the public networks so that they could compete against the state monopolies. It is also pushing for the elimination of all monopolies, privatization of state administrations, and the introduction of foreign competition. Rates in the EU countries are still twice those in the United States, and the European Commission regards this as detrimental to the competitiveness of European firms and European economies. Therefore it is pursuing a strong liberalization policy (Beltz 1996: 46–47). The United States has pushed ahead with its liberalization policy, and in early 1996, it passed legislation that will eliminate the final barriers to domestic competition. It has also continued to encourage American firms to bypass the higher priced foreign telecommunications administrations through devices such as the use of private lines, calling cards, and call-back systems (Beltz 1996: 46; Sampson 1996: 28–29).

The pattern of international liberalization varies throughout the world, but, as illustrated by this analysis, the direction is clear. Still, it would be foolish to expect completely open international markets and rigorous compliance with international liberalization agreements (Noam 1995; Mody et al. 1995, Sampson 1996; Ryan 1997; Xavier 1997). There are a variety of roots of the liberalization trend (which are elaborated on below). The increased popularity of strategies of deregulation, and even more the U.S. campaign for international liberalization, exerted pressure on states to accept more competitive telecommunications markets, but it is questionable whether either was the central driving force. What has probably been much more influential are the dramatic and constant technological changes that brought about new and competitive telecommunications services, additional modes of transmission, and overlaps with other important industries. The incentives and pressure for international competition that are embedded in these technologically wrought changes are tremendous. If U.S. firms had not been at the cutting edge starting in the 1970s, the pressure would almost certainly have come from other industrialized states.

CONCLUSION

The international telecommunications regime has constituted quite an effective body of international regulations since the 1860s, and most of its regulations have been directed at promoting order and openness in the international economy.

There have been both formal and informal norms and rules relating to jurisdictional rights and technical standards that have shaped states' and firms' behavior with regard to international telecommunications; and there has been a clear purpose underlying the rules—namely, facilitating the flow of communications among firms and countries. From the early days of the telegraph in the middle of the nineteenth century through the multimedia era of the late twentieth century, international commercial interests and governments have sought to assure the free flow of information. States almost always settled conflicts over equipment standards and distributions of frequencies and GSO slots when technical problems affected international commercial telecommunications. The body of telecommunications technical standards has, in fact, been central to the operation of the global economy. As noted above, international capitalism established a mandate for interconnection, and this has always been embedded in the international telecommunications regime.

States, however, did not see high prices for international telegraph, telephone and radio services as serious obstacles to the flow of communications until the 1980s. In fact, for many years a large majority of states supported high international rates because they used international revenues to subsidize domestic telecommunications systems. It is quite likely that, until the last two to three decades, the high rates did not reduce the volume of communications significantly, but with the multiplication of services, and modes of transmission and falling costs for all services this has changed. Pressured by their multinational corporations and concerned about their firms' international competitive advantage, the industrialized states have moved haltingly toward liberalization despite the efforts of interest groups tied to the traditional monopolies to slow the process. While firms' and states' adjustments to changing technology have undoubtedly had a major impact on the course of international telecommunications regulation, other factors that have affected regulatory trends are the nature of international capitalism and the international distribution of power. These latter two factors have been particularly relevant in recent decades when the increasing prominence of neoclassical economic theory, deregulation, and United States power have been prominent in shaping the evolution of the international telecommunications regime. These factors have probably not been as influential as technological change in shaping the international regime, but they do deserve highlighting.

A recent article commented that "Consumer demands and new technology are breaking down barriers that have protected telecommunications operators from domestic and foreign competition" (Beltz 1996: 47). What was happening on the technological front in the 1990s was reflective of a trend that goes back at least to the 1960s. Previous discussions have alluded to various technological changes, but it is important at this point to review the basic parameters of the technological juggernaut that has exerted such powerful pressure for liberalization.

First, trade in telecommunications services, and trade in services more generally, has experienced a dramatic rise in international economic importance. Trade in services today accounts for close to 25 percent of international trade (Sampson 1996: 24). The use of international telecommunications services is rising rapidly, and this is reflective of their tremendous importance to all firms in the global marketplace. Starting in the 1970s, large firms realized that their access to specialized and low-cost telecommunications services would have major impacts on their competitive success, and some of them were able to lease lines from the national monopolies (Cowhey 1995: 176–77, Rutkowski 1995: 229). Also, governments increasingly recognized that "Access to modern telecommunications services is an essential condition for success in the development of national economies and for international trade" (Sampson 1996: 19). Technological change created important incentives for both firms and states to undermine national monopolies and the rules of the international cartel.

Second, modern technology created a variety of new and improved telecommunications services that to an extent could be used interchangeably, and providers of these services were bound to enter into competition with each other. Perhaps of greater importance is that the varied services could be offered by different modes of transmission and by different industries. The lines between the services offered by wired telephone networks, cable networks, radiocommunications, satellites, video networks, and computing networks have been blurred. Relevant to this development, Michael Porter has commented that "Companies with networks having varying technological capabilities can be readily redeployed to offer different kinds of services. Thus, rivalry . . . is actually intense and is likely to remain so for the foreseeable future" (1992: 41). Of course, the interrelationships between these industries have been blurred in part because of the common use of digital communications across all of them. In the words of Peter Cowhey, in the digital age "technology has erased previous distinctions between equipment and applications" (1995: 179; Also Karlsson 1998: 2–5; Blackman 1997: 1; Noam 1995: 51–53). The increase in different means of transmission, of course, has not ended. Internet telephone has emerged recently, and in the words of The Economist it is "a time bomb" waiting to explode in the international telephone market (2 May 1998: 57; Also Beltz 1996: 48). In addition, low earth-orbiting (LEO) satellite systems that will permit telecommunications between any two points on the earth without going through hard wires will soon be in use. The Economist has noted that these new overlapping industries are extremely difficult to regulate and that vigorous competition is a natural byproduct (14 April 1998: 13; 16 May 1998: 17–18). On top of these developments, modern means of transmission such as fiber-optic cables, microwave, and satellites are increasing in capacity so rapidly that there are often incentives for firms to cut rates to increase market share (Porter 1992:

41). Looking at the massive technological changes over the past several decades, it is difficult to imagine that some firms and their home states would not have promoted competitive markets and consequently the demise of the traditional international cartel.

The distribution of power and the interests of the most powerful state(s) are always important factors in international politics. In the case of telecommunications, there is little evidence of marked divisions among the industrialized and even the non-industrialized states from the late nineteenth century through the 1960s. All states wanted technical interconnection and financially viable state monopolies. Starting in the 1970s, the United States and a few other states began to support those firms that called for some liberalization of market access and pricing (initially in the form of firms' leasing of lines), and with the breakup of AT&T in 1980, the United States embarked on a vigorous campaign to promote competitive international markets for telecommunications services. The United States became the central force in promoting liberalization accords in the GATT and WTO. It is, however, important to remember that the United Kingdom and Japan became active supporters of most U.S. policies starting in the early 1980s, and firms in all industrialized states soon began to pressure their governments to follow the lead of the liberalizing states. In fact, firms in many countries violated the cartel rules of the ITU since they knew that they would suffer seriously if they did not adopt more flexible competitive stances. Firms, in a sense, legislated the emergence of new rules by their own behavior. To quote one telecommunications expert: "recent trends highlight the growing importance of marketplace developments to public institutions in setting the de facto rules of the game" (Beltz, 1996: 49). Multinational firms would have found it much more difficult to break down the traditional cartel system without the backing of the U.S. government, but some governments would have eventually responded to the pressure of these firms if the United States had not done so (Aronson and Cowhey 1989; Cowhey 1995; Rutkowski 1995).

International telecommunications have developed since the middle of the nineteenth century, in a global economy dominated by capitalism. Through the first half of the twentieth century mercantilist sentiments, including states' concern about becoming dependent on foreign states for important national economic needs, supported national telecommunications monopolies and an intergovernmental cartel. The mercantilism of this era was, however, not of a highly autarchic and isolationist character. Capitalist states felt inclined to promote beneficial international economic exchanges and investment so long as they or their nationals maintained control of certain crucial industries. This concern for national control over the economy has certainly not disappeared today, but it is more muted. One reason for this is the popularity of neoclassical economic theory with its derivative prescription that states should promote competitive national and international markets. These prescriptions are often conflated with the strategy of deregulation. (Re-regulation would probably be a more appropriate term.) Deregulation achieved strong support in U.S. economic and policy circles in the 1970s, and it spread beyond the United States afterwards. A call for deregulation was, in fact, a kind of rallying cry in the political battles over the breakup of the monopolies and cartels in economic sectors such as telecommunications and air transport (Zacher with Sutton 1996; Karlsson 1998: 2–5).

While the popularity of deregulation certainly encouraged the process of liberalization in telecommunications, the key question is whether telecommunications would have taken the same course without the explosion in the literature on deregulation in the 1970s and 1980s (Breyer 1982). In other words, would liberalization have occurred if the capitalism of the 1950s and 1960s continued into subsequent decades? The argument here is that the recent changes in technology would have had similar liberalizing effects in the pre-deregulation age. The incentives for firms and some states to push for liberalization would not have been particularly different in the capitalism of the 1950s. The recent popularity of deregulation probably speeded the liberalization process, but liberalization was almost certain to occur.

Since the mid-nineteenth century, international capitalism has established the basic incentives for firms to promote international interconnection, and certain changes in its ideological orientation at least facilitated the movement toward liberal market access and competitive pricing starting around 1980. Likewise, the United States policy was very important in jump-starting international liberalization over the past two decades. On the other hand, it is difficult to imagine that other countries would not have eventually supported liberalization given the incentives for firms to promote foreign market access and lower telecommunications rates. The main motor for international political change in international telecommunications has been technological change and the evolving options it has presented to international firms.

A conclusion that can be drawn from the above analysis is that the dramatic developments in telecommunications or information technology have reduced the role of the state in national economies. States have judged that they should privatize their telecommunications administrations or should permit private competition to the state administrations. Influence over the character of telecommunications markets has fallen increasingly on private firms although states, of course, maintain certain regulatory roles. Still, there are limits to what the states judge that they should and can do. Technological change has quite simply created strong economic incentives to permit a significant degree of open competition and to encourage foreign investment—and consequently, to constrain the economic role and power of the state (For a broad perspective of this development, see Rosenau in this volume.)

REFERENCES

Aronson, Jonathan D., and Peter F. Cowhey. 1988. *When Countries Talk: International Trade in Telecommunication Services.* Cambridge, MA: Ballinger.

Bar, Francois, and Michael Borrus. 1987. *From Public Access to Private Connections: Network Policy to Private Connections.* Berkeley, CA: Berkeley Roundtable on the International Economy.

Beltz, Cynthia. 1996. Talk is Cheap. *Reason.* (August/September): 45-49.

Besen, S.M., and L.L. Johnson. 1986. *Compatibility Standards, Competition, and Innovation in the Broadcasting Industry.* Santa Monica, CA: Rand.

Blackman, Colin. 1997. Globalization, convergence, and regulation. *Telecommunication Policy.* 21: 1-2.

Bressand, Albert. 1988. Interconnection and the Think Net Process. *Interconnection: Vol. II.* Paris: Promethee. 3-6.

Breyer, Stephen. 1982. *Regulation and Its Reform.* Cambridge, MA: Harvard University Press.

Cerni, D.M. 1984. *Standards in Process: Foundations and Profiles of ISDA and OSI Studies.* Washington, D.C.: National Telecommunications and Information Administration, Department of Commerce.

Cerni, D.M., and E. M. Gray. 1983. *International Telecommunication Standards: Issues and Implications for the 80's.* Washington, D.C.: National Technical Information Service, Dept. of Commerce.

Chapuis, R. 1976. History of Regulations Governing the International Telephone Service. *CCITT Reprint Series.* Geneva: ITU

Christol, Carl Q. 1982. *The Modern International Law of Outer Space.* New York: Pergamon.

Codding Jr. G.A. 1972. *The International Telecommunication Union: An Experiment in International Cooperation.* New York: Arno Press.

Codding Jr., G.A., and A.M. Rutkowski. 1982. *The International Telecommunication Union in a Changing World.* Dedham, MA.: Artech House.

Comor, Edward A., ed. 1996. *The Global Political Economy of Communication: Hegeomony, Telecommunication, and the Interntional Economy.* New York: St. Martin's Press.

Cowhey, Peter. 1995. Building the Global Information Highway: Toll Booths, Construction Contracts, and Rules of the Road. In William Drake, ed., *The New Information Infrastructure: Strategies for U.S. Policy.* New York: Twentieth Century Fund Press. 175-204.

Drake, Willam J., and Eli M. Noam. 1997. The WTO Deal on Basic Telecommunications: Big Bang or Little Whimper? *Telecommunication Policy.* 21: 799-818.

Fjordbak, Sharon L. 1990. The International Direct Broadcast Satellite Controversy. *Journal of Air Law and Commerce*. (Summer): 55 903-38.

Fredebeul-Krein, Markus, and Andreas Freytag. 1997. Telecommunications and WTO Discipline: An Assessment of the WTO Agreement on Telecommunication Services. *Telecommunication Policy*. 21: 477-91.

Headrick, Daniel R. 1991. *The Invisible Weapon: Telecommunications and International Politics, 1851-1945*. New York: Oxford University Press.

Herring, J.M., and G.C. Gross. 1936. *Telecommunications Economics and Regulation*. New York: McGraw Hill.

Hills, Jill. 1986. *Deregulation Telecoms: Competition and Control in the United States, Japan, and Britain*. London: Pinter.

Horwitz, Robert Britt. 1989. *The Irony of Regulatory Reform: The Deregulation of American Telecommunications*. Oxford: Oxford University Press.

Hufbauer, Gary, and Erika Wada, eds. 1997. *Unfinished Business: Telecommunications After the Uruguay Round*. Washington, D.C.: Institute for International Economics.

Johnson, Elizabeth. 1986. Telecommunications Market Structure in the USA: The Effects of Deregulation and Divestiture. *Telecommunication Policy*. 10: 57-67.

Karlsson, Magnus. 1998. *The Liberalization of Telecommunications in Sweden: Technology and Regime Change from the 1960s to 1993*. Sweden: Linkoping University.

Leive, David. 1970. *International Telecommunications and International Law: The Regulation of the Radio Spectrum*. Dobbs Ferry, NY: Oceana.

McKnight, Lee, 1987. The International Standardization of Telecommunication Services and Equipment. In E.J. Mestmaecker, ed., *The Law and Economics of Transborder Telecommunications*. Baden-Baden: Nomos Verlag. 415-36.

Mody, Bella, Johannes M. Bauer, and Joseph D. Straubhhaar, eds. 1995. *Telecommunications Politics: Ownership and Control of the Information Highway in Developing Countries*. New Jersey: Lawrence Erlbaum Associates.

Mueller, Milton, and Dale Thompson. 2000. ICANN and INTELSAT: Global Communication Technologies and their Incorporation into International Regimes." Paper presented to the Annual Meeting of the International Studies Association, Los Angeles. (March)

Nicolaidis, Kalypso. 1995. International Trade in Information-Based Services: The Uruguay Round and Beyond. In William Drake, ed., *The New Information Infrastructure: Strategies for U.S. Policy*. New York: Twentieth Century Fund Press. 269-304.

Noam, Eli M. 1995. Beyond Telecommunications Liberalization: Past Performance, Present Hype, and Future Direction. In William Drake, ed., *The New Information*

Infrastructure: Strategies for U.S. Policy. New York: Twentieth Century Fund Press. 31–54.

OECD. 1988. Telecommunication Network. Paris: OECD.

Porter, Michael E. 1992. On Thinking About Deregulation and Competition. In Harvey M. Sapolsky, et al., eds, The Telecommunications Revolution: Past, Present, and Future. London: Routledge. 39–44.

Rutkowski, Anthony M. 1995. Multilateral Cooperation in Telecommunications: Implications of the Great Transformation. In William Drake, ed., The New Information Infrastructure: Strategies for U.S. Policy. New York: Twentieth Century Fund Press. 223–51.

Ryan, Daniel J. ed. 1997. Privatization and Competition in Telecommunications: International Developments. London: Praeger.

Sampson, Cezley I. 1996. Liberalisation of Trade in Telecommunications Services and the Implication of GATS/WTO for Developing Countries. Intermedia. (October/November): 24 19–30.

Savage, James D. 1989. The Politics of International Telecommunications. Boulder, CO: Westview.

Singh, J.P. 1999. Leapfrogging Development? The Political Economy of Telecommunications Restructuring. Albany, NY: State University of New York Press.

Tarjanne, Pekka J. 1990. Access to 1990, Telecom Networks and Markets. Transnational Data and Communication Report. (February)

The Economist (Weekly).

Wellenius, Bjorn et al., eds. 1989. Restructuring and Managing the Telecommunications Sector. Washington, DC: World Bank.

Xavier, Patrick. 1997. Australia's Post-July 1997 Telecommunications Regulation in International Context. Telecommunication Policy. 21: 533–51.

Zacher, Mark W., with Brent A. Sutton. 1996. Governing Global Networks: International Regimes for Transportation and Communications. Cambridge: Cambridge University Press.

CHAPTER NINE

UNDERSTANDING SHIFTS IN THE FORM AND SCOPE OF TELECOMMUNICATIONS GOVERNANCE: CANADA AND THE UNITED STATES IN THE TWENTIETH CENTURY

STEPHEN D. MCDOWELL

INTRODUCTION

The telecommunications sector has undergone a number of significant changes in the last decades. These include tremendous growth in business and consumer demand for communications services and information technologies, the building of cross-industry alliances and mergers and acquisitions among communications and information firms previously operating only in distinct sectors, the creation of transnational alliances and ownership relations among firms previously operating only in single countries, and the rapid introduction of new digital technologies at various points in communications networks. In order to facilitate, accommodate, and actively promote industry and market restructuring, governments have been called upon to undertake a wide number of new policy directions and governing practices.

In addition to new policies, more fundamental alterations in the ways in which telecommunications activities are governed have been introduced, as have changes in the actual organizational entities that are the locus of telecommunications governing activities. The chapter addresses the nature of these changes, and discusses different ways to understand why these changes in governance are occurring. It considers these changes in Canada and the United States, where the private sector has long been active in telecommunications, and where a variety of governing bodies and forms of governance have been in place throughout the twentieth century.

The numerous existing forms of telecommunications governance are accompanied by new public policy measures and directions, as well as by non-state forms of governance. Changes in the forms of governance have occurred along-side the reorganization of the geographic scope of telecommunications gover-nance, with significant activity taking place at the multilateral, regional, bilateral, and sub-national levels, in addition to traditional nation-state governance of telecommunications. This chapter outlines the changing role of solely national telecommunications governance, and sets this in the context of emerging pat-terns of supra-national and sub-national governance.

PRODUCTION, TECHNOLOGY AND GOVERNANCE

Two sets of developments are most often related to changes in governance; the or-ganization of economic production, and the development of new information and communication technologies. A critical political economy perspective argues that these changes in economic patterns, technology, and governance must be related to each other in a holistic, historical, and nonlinear fashion. New technologies have been designed and deployed to serve the purposes and projects of powerful organizations and social groups, such as states and transnational enterprises and their managers. The economic, cultural, and political patterns referred to as glob-alization also arise as a result of efforts of firms and states to compete in or more effectively control global markets, and their efforts along these lines lead to in-vestments in certain technical infrastructures and economic policies. Rather than assuming a certain type of technical change or economic restructuring as the starting point for exploring changes in governance, we also need to consider the origins and determinants of those broad sets of change.

The analysis in this chapter draws from historical and critical approaches in the study of international political economy and world order, especially that of Robert Cox (1987) and other neo-Gramscian political economists. The paper also draws insights regarding the spatial organization of production, governance, and social life and cultural meanings from economic and cultural geographers dealing with political economy and communications (Harvey 1990; Castells 1996; Gra-ham and Marvin 1996).

State power and forms of state should be seen in historical context, rather than abstract containers of sovereignty. Some states and forms of state will bene-fit from changes while others may lose out. Conventional accounts tend to assume that state organization, goals, and relationships with civil society remain similar, even if there are other new actors and even if there are new forms of power and threats. Rather, some states are undertaking efforts to benefit from new economic and technical patterns, and working to introduce new forms of governance that involve relations with non-state actors and supra and sub-state governing bodies. Although they may share the conceptual category of state, the capacities of small states and large states may differ to make them, in many respects, more different than similar in the global order.

Governance is chosen in this analysis over a number of other possible terms such as regulation, policy, or law. This term is selected for several reasons. Some terms, such as regulation, although they may refer to broad sets of state activities in the French regulation school, often are used to denote specific forms of policies or state intervention. Most often in telecommunications, regulation refers to the practice of price and entry regulation as practiced in the United States and Canada in the mid-twentieth century. Governance provides a way of thinking about activities such as legislation and regulation, while recognizing that these are part of a broader set of public activities. Secondly, the conceptual split between national law and international activities is often so great, that terms such as law or regulation are taken to mean very different things in the national or international context. Governance provides a way of relating and comparing these activities as part of a global order, without the discussion being overshadowed by national/international considerations. As a broad and somewhat abstract term, governance requires further specification in specific times and places. Hence, it encourages investigation of institutions and practices in particular times and places. As a flexible term, governance is also appropriate for thinking about institutional, economic, and policy change. As a term that can include both formal public policies as well as informal patterns of economic and social regulation, governance allows for the consideration of state and non-state forms of influencing social and economic relations.

Governance can be undertaken by public institutions, through laws, policies and programs that guide and regulate economic and social activities. This can include direct regulation of prices of services and the entry of firms into a market (at one end of a continuum, a long tradition in telecommunications), or simply setting and supporting rules for market exchanges and property relations or non-market activities, such as family law and rules governing the voluntary sector or industry self-regulation.

Work by cultural and economic geographers is especially useful in problematizing the spatial aspects of politics, culture and identity. This spatial aspect is also central in discussions of governance that may give rise to questions about sovereignty, since this is in part a territorial concept. To what extent do policy makers, business groups or various identity/membership groupings understand their concerns and actions to be set in a specific geographic context or scope? Or are these cosmopolitan ideas and understandings? The spatial conception of policy problems or identity/membership may overlap, and may not be exclusively assigned to one scope of governing body. Hence, in addition to the spatial organization of sectoral policy issues—in telecommunications, the main focus of this chapter—it would also be useful to consider reflections on the geographic and social places and spaces orienting cultural understandings and identity formation.

The neo-liberal state form of governance can be seen as a response and a contribution to globalization of economic processes and relations (Niosi 1991). The globalization of production, trade and investment has been formalized in new international trade and investment agreements, such as the Canada-U.S. Free Trade Agreement, the North American Free Trade Agreement and the Uruguay Round of negotiations in the General Agreement on Tariffs and Trade. Globalization and increased integration with international trade and investment in international political economy are also linked to state policies—such as deregulation, privatization, commodification of more social relationships—and to social changes in Canada and the United States. The liberal, outward-oriented state and the changing shape of international political economy are related to new forms of state intervention in domestic political economy. Measures in trade agreements significantly reduce the ability of governments to shape and direct economic production and exchange. Foreign companies must be treated similarly to national companies in an expanding number of sectors and there are few allowable restrictions on investment.

Neoliberal governance is also very concerned with the spatial organization of production, and how this relates to economic, political and social life. As a result of the emergence of clusters of production centers, national economic policies may be less appropriate than regional or even urban economic development policies. The connection between the geographic scope of a formal policy area and the mapping of relevant economic activity may be indistinct.

THE GOVERNANCE OF TELECOMMUNICATIONS

Telecommunications is defined for the purposes of this chapter as electronic communication at a distance, with capabilities allowing interactive information exchange and communications in real time or a relatively short time. The types of services this chapter is concerned with are those offered through public common carrier networks, services available to the general public. While this definition includes voice and data communication, it is less relevant to traditional mass media and cultural expressions, such as radio and television broadcasting or the distribution of films and sound recordings.

Robert Babe's study of telecommunications in Canada is especially helpful in tackling the assumptions and claims of a technologically determined view of telecommunications policy (Babe 1990). It examines the interrelationship between industry structure, government policy, and the uses of communications technology in Canada. Such an approach, therefore, might note the importance of examining how the growth of patterns of regional integration (or the shifting scope of communications policy institutions) is related to changes in industry structure, and to the organization and use of communications technologies. One of Robert Babe's central arguments is that Canadian telecommunications policy, regulatory patterns, and industry changes significantly influence, and are influenced by, the uses and control of technology (but not driven by technical dynamics, as is often claimed in telecommunications policy literature). The (monopoly, oligopoly or competitive) structure of the telecommunications industry, the historical forms of government intervention, and the design and uses of telecommunications technology must be examined together.

Telecommunications governance intervenes in, is set in, and is influenced and shaped by the organization of production and by the industry structure, and by technology developments and uses. This might be seen as the terrain or context of telecommunications policy and governance, and the linkages between forms of governance and these broader developments are explored in the second part of the chapter.

CHANGING FORMS AND NON-STATE GOVERNANCE

The key policy debates in telecommunications governance over the past two decades have concerned what the goals of public bodies and agencies should be, and what forms of intervention are appropriate. Terms such as deregulation, liberalization, and technical convergence captured and held a preeminent position in much of the debate over appropriate forms of communications governance.

Beginning the 1980s, for instance, there were pressures in numerous countries for changes in forms of regulation of the telecommunications industry (such as deregulation) and to allow international provision of telecommunications services through more open trade and investment. A number of policy measures have been implemented. These include measures such as the introduction of (limited) competition among telecommunications service suppliers, more trade and foreign direct investment in telecommunications service and equipment production, the privatization of state companies, and mergers among different sectors of the telecommunications industry. Also notable are changes such as the combined use of different communications technologies in the concept of a multimedia information superhighway, the implementation of forms of regulation that allow providers more price flexibility, more rapid introduction of new services, and greater public policy focus on the use of telecommunications for social and economic development (Kahin and Wilson 1997; Kahin and Nesson 1997). However, a variety of forms of public intervention aimed at achieving social, economic, and political goals by directing the design and deployment of communication technologies are also being undertaken by public bodies alongside deregulation (Mosco 1988). These goals and programs might include promoting effective government communications and information services, ensuring privacy protection and information security, supporting research in communications technology, planning infrastructure investments, or facilitating exports, strategic partnering or inward or outward foreign direct investment.

An emerging acceptance and institutionalization of forms of governance targeted at, and limited to, supporting market economic interactions is illustrated by these goals and programs. What is less clear is whether forms of governance are likely to emerge to promote equity objectives. These examples also demonstrate that, even alongside shifting forms of regulation, deregulation, and liberalization, governments do retain, and are using, a variety of forms of policy intervention. However, some of these forms of governance represent a significant contraction in the role of the state.

A problem at the core of governance is the extent to which formal state patterns of governance interact with non-state ordering of social relations and political and economic organization in civil society. The widespread use of terms such as society, civil society, social institutions, the market, self-regulation, and the recognition of the importance of organizations such as churches, professional association, trade association, and nongovernmental organization all indicate the importance of non-state bodies in ordering or governing a variety of social and economic activities at the national and international level.

One conscious and direct method of control of behavior is self-regulation by groups. It has been used by professionals and by industry sectors at different times as a mechanism to resolve possible political problems and to preempt the perceived need for public governance. Often working alongside public licensing bodies, certain groups, such as doctors, lawyers, and engineers are mandated and expected to regulate their own members' behavior and services to the public. Self-regulation is often sanctioned or sought by the state, in that state managers and the public perceive a significant lightening of the volume and burden of administrative actions and decisions as a result of professional self-regulation.

For instance, technology standard-setting now involves extensive industry participation in North America and Europe. In some cases, such as the digital television technology, the public sector has declined to set any standard, even after extensive negotiations among the industry and the public sector. In telecommunications network standard-setting, industry representatives may be formally recognized at the national level through self-regulatory bodies, through consultation in public sector standard-setting exercises, or through participation in national delegations or industry groups at the international level. Therefore, in considering the emerging forms of telecommunications governance, procedures and practices wherein public governance is significantly constrained or delegated to self-regulatory bodies should be examined closely (Cutler et al 1999; Vogel 1996).

Self-regulation is mode of governance that has been increasingly proposed for communications networks. The V-Chip technology in the United States depends on the use of a ratings system that is designed largely by industry groups and will be implemented by industry groups. Concerns about children's access to certain types of internet content, initially met by an attempt to include the so-called "Communications Decency Act" in the United States Telecommunications Act of 1996, are now being responded to by suggestions that services be labelled and regulated by the industry. Governance of the Internet has been presented by some as an example of spontaneous or conscious self-regulation.

Although industry self-regulation and participation in technical standard-setting are not central themes in the study of international politics, another set of actors in of non-state governance—non-governmental organizations—have been widely recognized as important actors in the post-1945 system of international organization and governance. As Ness and Brechin (1988: 250) note, the United Nations has the capacity "to grant formal recognition to nongovernmental organization." Similarly, studies of international civil society and its relation to forms of state and international organizations also highlight the importance of non-state bodies and social linkages in the form and practice of governance.

The variety of non-state organizations and the range of participants in forms of telecommunications governance pose several conceptual issues and theoretic questions for political economy and governance. Despite an acceptance of a growing role for non-state corporate and civil society groups and organizations, neo-realist international theorists still see the defining characteristic of the modern world as being the preeminent role of states and the interstate system as determining specific outcomes and as defining the unique and essential characteristics of the global system. Pluralist and critical approaches use terms such as global governance or world order to explore the ways in which world politics are developing and changing without deciding *a priori* that the state and the interstate system remain the unchanging centerpiece of world politics.

The consideration of non-state forms of governance introduces the possibility of unlike units in the field of analysis. This complicates the study of international or world politics, which is premised on the existence and primary important of the interstate system. The problems encountered are significant in trying to deal, not just with differing forms of telecommunications governance by states, but also with a configuring of governing institutions composed of unlike units that cannot be given, a priori, any set ordering of importance.

Several theorists have looked to politics and society prior to the modern state system to begin to conceptualize how the emerging system and global society might be organized. Hedley Bull has called this a "neo-medievalism", what Barry Buzan calls a "label for a system of unlike units." In his discussion of international society, Buzan argues:

there is no logical reason why neomedieval versions of anarchic systems could not develop international societies. In such a system, shared identity as a similar type of unit is by definition not a basis for society. In a medieval international system, the only possibility for shared identity is not acceptance of likeness of units but in acceptance of a set of rules that legitimize responsibilities among functionally differentiated actors. Compared with the primitive possibility of shared identity among like units, this is a complex and sophisticated form of international society. It is difficult to examine such an arrangement coming about from scratch. . . . Nevertheless, the neomedieval form is worth keeping in mind as an evolutionary possibility for highly developed international societies (1993: 335–336).

As quoted in Fischer, John Gerard Ruggie argues, "The feudal state, if the concept makes any sense at all, consisted of chains of lord-vassal relationships. Its basis was the fief, which was an amalgam of conditional property and private authority. Property was conditional in that it carried with it explicit social obligations . . . Moreover the prevailing concept of usufructure meant that multiple ties to the same landed property were the norm. As a result, the medieval system of rule reflected 'a patchwork of overlapping and incomplete rights of government' which were 'inextricably superimposed and tangled,' and in which 'different juridical instances were geographically interwoven and stratified, and plural allegiances, asymmetrical suzerainties and anomalous enclaves abounded'" (1992: 432).

More recently, Stephen Kobrin (1998) addressed this concept directly in a discussion of telecommunications governance by linking "neomedievalism and the postmodern digital world economy." In this period of "turbulent, systemic change in the organization of the world economic and political order," Kobrin argues that the "medieval metaphor should be seen as an inter-temporal analog of comparative political analysis. It allows us to overcome the inertia imposed by our immersion in the present and think about other possible modes of political and economic organization." Kobrin's discussion of this metaphor in the context of contemporary political economy points to a number of elements, including geographic space and borders, the ambiguity of authority, multiple loyalties, transnational elites, shifting distinctions between public and private property, unifying belief systems, and supranational centralization.

The use of the medieval case as an example of a specific historical form of governance, is introduced here, as Kobrin suggests, as a conceptual frame that can assist in forming questions and thinking about forms of governance of telecommunications that may be emerging in the early twenty-first century. Specifically, to what extent do patterns such as industry self-regulation and self-governing communications user organizations represent the introduction of unlike units as complements or collaborators alongside state and public bodies in telecommunications governance? What theoretic and historical importance will these new forms and new configurations of various modalities have for telecommunications governance?

THE SCOPE OF TELECOMMUNICATIONS GOVERNANCE

A second question this chapter will explore regarding the new telecommunications policy environment is the possibility of a reconfiguration of the spatial scope of telecommunications governing bodies. The scope of jurisdiction of communications governance refers to the geographic range of particular public organizations in formulating telecommunications policy and in using telecommunications technology to promote social and economic development. Although the power of states and their rights of sovereignty are often claimed to be indivisible in the modern period, in practice telecommunications governance in the twentieth century has moved through a variety of orders. As is argued below, some, such as the ITU regime from the nineteenth century to the 1980s, were very supportive of the nation-state competence. In a federal system, governing responsibilities and jurisdiction are sometimes divided among different levels of government, or in many cases shared among different levels of government. Apart from any abstract models, public policy and the overall role of states evolve historically, a result of struggles at specific times and places among different social groups and institutional actors. The discussion below emphasizes the importance of telecommunications activities of cities, states and provinces in federal systems of governance, nation-state agencies, regional trade agreements and organizations, and multilateral organizations (Aronson, Diebert, both in this volume).

Sub-national units, such as cities, states or provinces, and sub-national regions, increasingly are seen by businesses and economic planners as possessing the appropriate geographic scope for organizing certain industrial infrastructure and economic activities, and responding through state policy, planning or coordinating nongovernmental social organization (Brunn and Leinbach 1991; Drache and Gertler 1991; Kellerman 1993; Pool 1990). Cities or urban conglomerations are, in the view of some economic geographers, the key unit for understanding economic and social development, rather than making reference solely to national economies or national development.

While conflicts in federal systems of governance are most often perceived as legal and juridicial questions to be decided by constitutional interpretation by the courts and by precedent, a political economy analysis would suggest that the types of issues that emerge in state-federal government debates, and the roles which each level of government is called upon to pursue, should be related to economic organization and to the uses of technologies. James Carey (1988), for instance, argues that among the implications of the development and use of the telegraph in the nineteenth century was the creation of regional and then national markets in the United States. As well, firms began to manage production and marketing on a national basis (Beniger 1986).

Regional integration has generally been understood as a shift of state functions, especially for particular policy areas, to suprastate organizations composed of geographically contiguous nation-states. Both as an historical process and an area of theoretical reflection, regional integration poses a number of difficulties for thinking about nation-states, international political economy and international organizations. Contemporary political institutions (both the nation-state and the global international organization of economic activities and policy areas) are sometimes seen to be under challenge both from new and longstanding regional institutions and bodies. In the post-1945 period, while there was much investigation of the possibility of the formation of a European security and economic community (Pentland 1973), international organizations operating at a global level were seen, in a functionalist sense, to have the scope of competence more appropriate to deal with specialized economic relations and technical coordination (i.e., finance, trade, transportation, communications).

Paul Taylor's discussion of regionalism relates its usage to functionalist/ sectoral and to geographical approaches to understanding the role of international organization as a process in world politics. Sectoral approaches to regionalism are "concerned with individual problems and hold that they can be solved in their own terms" by focusing on an issue area or problem (Taylor 1993: 7). A sectoral approach can be contrasted with geographical understandings (which focus on the appropriate spatial dimensions of problems, institutions or political projects) of political economy: "One recurrent theme, however, is a concern with that particular scale of geographical area which is best fitted to the performance of tasks judged crucial for the welfare of individuals, or for the advantage of governments" (Taylor 1993: 7). Taylor argues that sectoralism has emerged as the main conceptual competitor to regional understandings of international organization, in part because the spatial "alternatives to regionalism [have] an extreme form—the exclusive nation state or the organic globe" (1993: 8). Taylor's discussion is useful in highlighting the conceptual difficulties encountered in the present study.

Again, the neo-medieval metaphor illuminates the possibility that the relationship between governing bodies with different geographic scopes should not be seen only as being competitive (an either/or of global versus regional versus national jurisdiction). These roles may be seen as complementary pieces of an overall global institutional order. These arrangements might include, for instance, state institutions and policies supporting international institutions and practices at a regional and global level, or networks of bilateral relations among states. While regional or international influences are often seen as offering invasions or threats to national sovereignty—whether seen as the control of borders, national decision processes, and a central and shared identity or public space—it is important to note that international rules and institutions may both empower and limit powerful state and non-state actors and may define the sphere of state action in particular issue areas. They can serve not only to enhance the scope of state action and the achievement of particular interests, but also to constitute the sphere of national and state decision making. If state sovereignty is historically constituted by shared practices and beliefs in international relations, then international governance can be seen as more than a strict subsidiary of state power or as a competitor for state power. To draw out this analysis, the overall structure of the relationships among sub-state, nation-state, regional, and international organizations involved in communications policy should be considered as contributing to or constituting the role and responsibilities of each particular policy body.

TELECOMMUNICATIONS GOVERNANCE IN NORTH AMERICA

Can the spatial reorganization of communications policy institutions—across the sub-national, national, and international levels of policy change—be connected with a general contraction in the policy role of various public bodies in certain areas (i.e., less commitment to universal service and greater discussion of the use of telecommunications to enhance economic development)? The historical account here argues that the reorganization of national forms of telecommunications regulation and policy in contemporary policy is consistent with increasing regional integration and global coordination in telecommunications policy. Hence, the historical changes in the form of telecommunications governance and the scope of telecommunications governance may be understood as arising from similar strategies or dynamics.

This section provides historical evidence to support the claim that there is an ongoing recasting—or perhaps a diminishing—over the late twentieth century of the sole governance of telecommunications by nation-states. Telecommunications governance was formulated in the late nineteenth century in an environment of European imperial rivalry. In the mid-twentieth century, governing practices reflected the preeminence of national goals and purposes, and were determined primarily by nation-state institutions. International organizations served to coordinate networks while respecting and reinforcing this aspect of nation-state sovereignty. In the 1980s and 1990s, new multilateral, regional, and bilateral trade agreements and commitments reflected a surge of institution building at the supra-state level. The primary role of the national government has been recast by greater state commitments to continental bodies in North America, and to multilateral agreements in which national governments have agreed to constrain certain types of policy measures. At the same time, other public organizations have began to deal more extensively with telecommunications policy issues, including cities, counties, regional development authorities, and provinces or states (at the sub-national level). States and city/local governments are paying greater attention to telecommunications infrastructure planning and deployment, and to the role of telecommunications in social and economic development.

COMPETITION AND CONSOLIDATION IN THE LATE NINETEENTH CENTURY

In the telecommunications sectors in the late nineteenth century, there was extensive private investment in international telegraph cable construction, and this was set in the context of imperial competition among European powers and U.S.-Europe competition in Central and South America. According to Graham Thompson (1990), Canada acted as a subimperial power in the British empire in international telecommunications networks, even though the Canadian telephone industry was owned by American companies (see also Fortner 1979). Sir Sanford Fleming took a lead role in constructing an undersea telegraph cable from British Columbia to Australia (for which he preferred public ownership), which linked the farthest parts of the British empire. As well, Canadians also invested in telecommunications and electricity companies in central and South America (Armstrong and Nelles 1988).

Similarly, a United States company—the International Ocean Telegraph company—sought to construct cable connections between Brazil and the islands of the West Indies. However, a monopoly over cable landing rights in Brazil was won by a British consortium, the Western Telegraph Company. The Mexican Telegraph Company (a New York company) constructed cables from Texas to points throughout Mexico in 1878, and its promoter—a Mr. Scrymser—in 1882 financed a cable running through Central America and South America to Lima, Peru. These cables later became part of All-American Cables Inc. One American commentator, Elihu Root Jr., viewed the competition among British and United States cable companies as "a death struggle on for the control of the South American communications situation" (Tribolet 1929: 42–45). Leslie Bennett Tribolet noted that "the British claim that the United States has an iron-clad communications monopoly in Mexico at present," and that "British political control in the various regions of the earth seems to coincide almost in direct ratio to the extent of the communications and propaganda monopolies in those same regions"

(1929: 45–46). This imperial rivalry conducted by proxy companies was at one time complicated by the possibility of a cooperative arrangement between Western Union of the United States and the British company (Western Telegraph Company), two of the four companies competing in South America for the development of communications (Tribolet 1929: 60). One partner, the Radio Corporation of America, also appointed a chair who held an effective veto, "thus carrying the principle of the Monroe Doctrine into the field of communications in the Western Hemisphere and giving the Americans effective leadership" (Ibid.:61).

The International Union of American Republics (now called the Organization of American States) was founded in 1890. Canada was not part of this organization, which consisted of the United States, Mexico and other Latin American and South American countries. The dominant position of the United States in the region and its claims of a sphere of influence under the 1820 Monroe Doctrine (and the practices of annexing territory, invading countries whose administrations pursued their own objectives, and the support of oppressive governments favorable to the Americans, which typified United States foreign policy) mitigated against the OAS as being a regional international organization presenting a strong voice independent of the United States's purposes.

While participating in the OAS, the United States did not join in all of the relevant treaties for international telecommunications (see Zacher, this volume). Along with Canada, the United States was not a signatory to the International Telegraph Convention of St. Petersburg of 1875, which reorganized the International Telegraph Union. George Codding states that the reason for the lack of direct participation in the Telegraph Union was that the United States's telegraph lines were privately owned, "a private enterprise not subject to the control of government," rather than nationalized (Codding 1952: 42). This policy arose "because of opposition of American companies to what they consider undue international control and regulation" (Tribolet 1929: 10–11). The United States did attend the St. Petersburg convention as an observer delegation. George Codding cites one United States participant in the 1875 conference who reflected: "The interests of the public who use the telegraph seemed to be entirely subordinated to the interests of the state and to the administrations; that is, to a fear lest improvement might produce less revenue than is got at present, and lest it might throw more work on the telegraph bureau" (1952: 65). However, American companies attended subsequent Telegraph Union conferences, as "they found it expedient to accept the International Telegraph Regulations in their international communications" (Codding 1952: 43).

The United States did sign the International Radiotelegraph Convention of London (1912), with the reservation that "its government is under the necessity of abstaining from all action with regard to rates, because the transmission of radiotelegrams as well as of ordinary telegrams is carried on, wholly or in part, by commercial and private companies" (Tribolet 1929: 21). In the 1920s, the United States and Canada both objected to proposals which would have linked the International Telegraph Union and the activities of International Radiotelegraph Conventions at the 1930 Brussels International Telegraph conference. The United States also pursued this policy in the Inter-American Committee on Electrical Communications in 1924, despite the opposition of all other countries except Argentina (Tribolet 1929: 22–23).

Several developments during this period are instructive. Firstly, as Robert Babe argues, the development of the industry structure followed as much from the use of financing strategies and the goal of obtaining monopoly licenses as it did from the dynamics of technological innovation. The level of *continental integration* between the United States and Mexico through telegraph combines was high (with Canada eventually building national telegraph networks controlled by the railways). United States companies' investment in Central and South America was set in the context of imperial rivalry, a contest the isolationist United States did not abstain from in the Americas. In the telephone industry, the Canadian and United States systems were very closely tied together through the Bell company. Secondly, these industry developments took place while the public policy mechanisms for guiding the development of national telecommunications systems were still weak. In Canada, the federal government supported the Bell telephone company by defending it from municipalities and provinces, only developing national regulatory legislation in 1917 in response to much populist protest. In the United States, debates about the division of governance in federalism have a long history, with disputes since the late nineteenth century centering around the role of the federal and state governments in regulating commerce, transportation, and financial services (Horwitz 1989; Teske 1994). Thirdly, the interlocking relationships between national industry structures and policy institutions, and among national institutional patterns and regional organizations and multilateral forums were already forming in the late nineteenth century. Canadians pursued cable construction as a sub-imperial power in the British empire, while increasingly linked into the growing North American telephone network. The United States noted in international conferences that telegraph rates were a private matter, and resistance by private sector companies to state intervention prevented the United States and Canada from joining the International Telegraph Union. It is significant, however, that the International Telegraph Union did have provision for private sector operators's participation, an exceptional limitation on the scope of state representation of national interests in international bodies (where states are often the only participants and sole legitimate representatives of industry groups in their territory).

STRONGER NATIONAL AND REGIONAL GOVERNANCE IN THE MID-TWENTIETH CENTURY

From 1920s to the 1970s a stable national/monopoly governing order developed in Canada and the United States. This was typified by greater national ownership of monopoly service providers or a shifting mix of public and private ownership, national equipment providers, public interest regulation of rates and services, and strict distinctions among different communications activities (Mosco 1988). Key communications policy issues that raise questions of the federal division of responsibilities in the United States include the consideration of technologies used for local and long distance services, accounting conventions for charging for the use of the network by numerous different types of services and users, and the appropriate role of state and federal governments in regulating intra-state and inter-state services. Telecommunications governance in the mid-twentieth century institutionalized the relationships among different levels of government, with representatives of the National Association of Regulatory Commissioners (state level regulators) participating in numerous joint boards with the Federal Communications Commission.

In this period, international traffic was coordinated through interconnection agreements, with less competition among international providers. Extensive work programs of global committees of the International Telecommunication Union negotiated technical standards (Codding and Rutkowski 1982). The Organization of American States also began forming a series of radio conventions in the 1930s. The first four Inter-American Radio Conferences of the OAS were held in 1937, 1940, 1945, and 1949. At the First Inter-American Radio Conference in Havana, November 1–December 3, 1937, accords were signed to explore setting up an office (Inter-American Radio Conference 1937). Canada was represented at this meeting by a two-person delegation. The problem of sovereignty and international representation was also raised. It was agreed that one vote should be accorded in the meeting to each government or body that met the qualifications of having a permanent population, a defined territory, a government, and capacity to enter into relations with other states. Colonies or administered territories would thus be excluded from voting, and other bodies that did not fit these criteria could have a voice at meetings, but not a vote. In addition to questions about the allocation of radio-magnetic frequencies and the suppression of interference among signals, one key area of discussion concerned the press use of telecommunications at low rates and with open access. It was resolved that "such economies be extended to and embrace all forms of telecommunications, including all methods of transmission now in use, such as telegraph, telephone, cable and point-to-point radiocommunication, including image, telephoto and facsimile, as well as proper forms hereafter to be developed, such, for example, as television" (Ibid.: 4).

As George Codding notes, the meeting illustrated a significant extent of regional coordination: "[c]oming as it did one year in advance of the next scheduled International Radio Conference, the Inter-American Radio Conference gave the countries of the Americas the chance to achieve a degree of solidarity with which to enter discussions at Cairo [in 1938] that no other region, or group of countries, could claim at that time" (Codding 1952: 160).

The third meeting of the Inter-American Radio Conference was held in Rio de Janeiro from September 3 to 25, 1945. It resulted in the conclusion of an Inter-American Telecommunication Convention on September 27, 1945 (Inter-American Radio Conference 1945). The signatories of the convention included Argentina, Bolivia, Brazil, Canada, Chile, Columbia, Costa Rica, Cuba, Dominican Republic, Ecuador, El Salvador, Guatemala, Haiti, Honduras, Mexico, Nicaragua, Panama, Paraguay, Peru, United States of America, Uruguay and Venezuela. Observers from Newfoundland, the Bahamas, and the British Caribbean colonies were also present. The convention recognized that "the public has a right to use public telecommunications services. The service, rates and guarantees shall be equal for all users in each category . . . without any priority and preferences" (Inter-American Telecommunications Convention 1945: 20). Rates for international services would be "fair, reasonable and equitable, and shall correspond to the service actually rendered." These pricing agreements and the priority placed on the open use of telecommunications by news services reflected to only non-technical aspects of the discussions, even in the context of the end of World War II.

In 1962, a proposal was discussed for the establishment of an Inter-American Telecommunications Commission (CITEL), which would "be responsible for promoting and facilitating Inter-American cooperation in the field of telecommunications [and] be a specialized inter-American organization" (OAS 1974: 1–2). This organization finally held its first meeting in 1965. Meetings of the Inter-American Telecommunications Commission (CITEL) continued, and with the sixth meeting of CITEL held in Caracas, Venezuela on September 5–6, 1971. This also coincided with the First Inter-American Telecommunications Conference held in the same city from September 5–11, 1971. This meeting, therefore, represented a significant reinvigoration of regional efforts in telecommunications. While Canada was not a member of the OAS or CITEL, it participated as an observer at these meetings at least since the 1970s.

Almost all of the resolutions of the 1971 meetings relate to technical matters and issues of system operations and coordination. What is also notable about the resolutions of the 1971 meeting is the focus on indigenous manufacture of "equipment, supplies and components for telecommunications systems in Latin America" (OAS 1974: 63–65). Resolution CITEL-I/47-71 noted that to promote and organize the industrialization of Latin American countries, to coordinate network development with equipment production, to reduce the dependence on foreign technology and large expenditures on imported equipment, and the promotion of the "fast growth of national networks in the years ahead justifies the establishment of new plants with potential economic benefits, and failure to do so would further increase the degree of technological dependency." The resolution argued that "the development of national manufacture of telecommunications equipment favors the attainment of national objectives and interest[s] of the Latin American countries" and would permit "comprehensive planning of the telecommunications sector," favor the adoption of more complex technologies and the beginning of applied technological research, foster investment and savings of foreign exchange by import substitution, and more balanced dealings with foreign suppliers.

This focus on technical issues and network development was also present in the 1975 meeting (OAS 1975). That meeting recommended the drafting of an Inter-American Telecommunications Development Plan, "to give countries of the region a framework that will incorporate in their development plans timely and adequate measures that will permit an efficient, economical, and coordinated flow of telecommunications traffic between countries of the region." This proposal was premised upon the view that countries needed a regional network, that the regional network would be an extension of the national development plans of each country, and that each country should specify and contribute its intra-regional objectives. In the 1960s, the focus of CITEL was on efforts to create an Inter-American Telecommunications Network, "the result [of which] was the establishment of a telecommunications infrastructure based on microwave networks. . . . CITEL concentrated also [sic] its efforts on rate setting and exploitation aspects and the study of spatial communications as a complement to the infrastructure began to take on importance" (OAS 1987: 7–8).

The telecommunications sector was somewhat unique in comparison to wider patterns of state-industry relations in the political economy of Canada and the United States. The state maintained a strong role as long as telecommunications were seen as public utilities. This uniqueness was especially marked in this sector, given the growing internationalism of corporations in the post-1945 period, which saw greater continental integration in manufacturing. Telecommunication was a national monopoly industry which was regulated by the state (or owned by public bodies in some cases). In an era of mass production of industrial and consumer goods, telecommunications was often seen as utility, providing services for the wealth-creating sectors of the economy. Technical and engineering problems rather than market and economic issues were the main questions in academic and policy enquiry.

Although supportive of open trade and investment in other sectors, the United States was able to support a strong role for regional organizations in planning infrastructure investment and development, as well as a strong role for the nation state in promoting industrialization. This policy was consistent with the regulatory and policy order at the national level in the United States, and was maintained despite the presence of dominant equipment export companies in the United States. Regional coordination and technical integration was consistent with—rather than in competition with—a strong role for national governance in telecommunications network planning.

The first reflection emerged in 1979 of the more political issues that were being discussed in other international organizations in the late 1970s, and the breakdown of the historical relationship between the role of states in national telecommunications and strong regional international organizations (OAS 1979: 6).

LIBERALIZATION AND REDEFINED REGIONALIZATION IN THE 1980S AND 1990S

During the 1980s, several processes began to significantly restructure telecommunications policy, planning and regulation. These took place at the national level in many nation-states, at the global level, as well as in the patterns of regional cooperation and coordination in Canada, the United States, and the Americas. The development of more regional governance activities in North America occurred alongside the development of multilateral agreements in the WTO and ITU. In addition to developing rules to facilitate international trade in telecommunications services and information technology, states were also making commitments to changes in national telecommunications rules and regulations, and to limitations of foreign ownership of telecommunications service providers.

In Canada, telecommunications governance, although set in a federal system, is seen, since the Supreme Court decision in 1987 of Regina v. AGT, to reside solely with the central government. The federal parliament, government, and its agencies were particularly active in communications the 1990s, enacting new telecommunications legislation, undertaking a review of telecommunications regulatory structure and pricing in the CRTC (Canadian Radio-television and Telecommunications Commission), as well as promoting the development and uses of new communications infrastructure and services. At the same time, other levels of government in Canada have considered the role of communications technologies and services in social and economic development, and have taken their own initiatives to advance the use of telecommunications. These activities suggest that a solely national program of communications regulation and policy is not seen as appropriate to meet the communications and development needs of all parts of the country. Although the federal organization of communications regulatory authority differs between Canada and the United States, the analytic literature and the actions of various levels of government suggest that questions about the fit between various scopes of governance are still relevant to the Canadian situation.

The liberalization in the United States began in the early 1980s with national initiatives (divestiture of AT&T, competition in long distance telephone service). In Canada, efforts to promote more open trade in informatics and telecommunications services in the Canada-U.S. Free Trade Agreement preceded national efforts, and were used to prompt efforts at national regulatory restructuring in the early 1990s. These processes continued with the signing of the North American Free Trade Agreement (NAFTA) which included Mexico, the United States, and Canada. Protection of basic telecommunications services for national carriers continued in the NAFTA agreement. Mexican barriers to trade in enhanced services were eliminated, and access to contracts offered by government telecommunications services in Mexico was also granted. Access to and use of telecommunications networks in the three countries is guaranteed. Equipment standards would be made compatible over time. Mexico would immediately remove tariffs on most telecommunications equipment, and phase out other tariffs over five years (Canada 1992).

The United States also undertook a number of initiatives over the late 1980s and early 1990s designed to encourage economic and telecommunications development in the Central American region. Beginning in 1982, the United States Telecommunications Training Institute offered free courses on various technical specialities (as did Canada through its own institution, for which it also provided OAS scholarships). The Caribbean Basin Initiative in the 1980s gave tariff concessions and access to the United States market for Caribbean countries. United States President George Bush's *Enterprise of the Americas* program was planned to encourage economic growth and economic policy reform. A cooperative arrangement between the OAS secretariat and the U.S. Trade and Development Program also led to the preparation in the late 1980s of a study on the modernization of the telecommunications systems and companies in five CITEL countries (OAS 1987: 5).

The activities of the Inter-American Telecommunications Commission (CITEL) of the OAS grew in the 1980s and 1990s, as did their importance in the view of industry actors and policy-makers in the United States and Canada. However, the organization was, according to at least one major member country administration, hampered by resource and organizational problems. These were already noted in the 1980s, but were exacerbated in 1991 "when the OAS decided to suspend funding for several activities of the Organization, including specialized conferences, such as CITEL. . . . Fortunately, some CITEL member countries, such as Canada, the United States, Mexico, and Uruguay, funded meetings held in their countries, providing their own funds for items that were usually covered with funds allocated by the OAS to CITEL" (OAS 1991a, 2).

A special effort was made in the late 1980s to build relationships between CITEL and many public and private organizations. It was reported in 1987 that

"Relations have been established with the following organizations: the Asociacion Hispanoamericana de Centros de Investigacion y Estudio de Telecomunicaciones (AHCIET), the Asociacion de Empresas de Telecomunicaciones of the Andean Sub-regional Pact (ASETA), the Comision de Telecomunicaciones de Centro America (COMTELA), the International Amateur Radio Union (IARU), the International Satellite Organization (INTELSAT), the Pacific Telecommunciation Council (PTC), and the International Telecommunication Union (ITU), with the closest possible relations existing in the case of the latter" (OAS 1987: 2–3). A resolution (COM/CITEL RES. III (XVII-86)) recommended the participation of public telecommunications service operating companies in the work of Permanent Technical Committee I as special guests, with no limit to the number of companies that might participate. They could participate in all activities except voting, and would be expected to make voluntary contributions to cover additional costs of participation. This direction was also confirmed in 1991, which noted increasing links between CITEL's activities and private telecommunications companies (OAS 1991b: 2). CITEL would aim to "strengthen the required interaction between CITEL and all the subregional and non-regional entities that are also engaged in activities aimed at the development of regional telecommunications, and, as a necessary step for CITEL to become coordinator of this development" (OAS 1991b: 2).

Many of the activities of CITEL during this period, following a decision of CITEL V, were to prepare common positions for ITU meetings: "to concentrate its efforts for this period on activities to achieve the coordination and harmonization of the region's countries in regard to technical criteria and standards, planning criteria and methods when in order, regulatory aspects, and so forth. It[s] aim was to arrive at common viewpoints and stands that would take account of this region's needs and better equip it to defend our interests successfully in other international fora with a view basically to having decisions adopted contribute to the development of telecommunication in the Americas" (OAS 1987, 3).

After 1994, CITEL was made up of the CITEL Assembly, which is the highest body in the organization, the Permanent Executive Committee, the Permanent Consultative Committees, and the Secretariat. The participants in CITEL meetings include delegations from OAS member countries, from states that are permanent observers in the OAS, observers from other OAS organs or Inter-American regional bodies, and observers from the United Nations and its specialized agencies. As well, subject to the approval of the Permanent Executive Committee of CITEL (COM/CITEL), observers may also include "international, regional, subregional and national agencies who have asked to participate in the meeting" (OAS/CITEL 1994b, 9–10).

In addition to the CITEL members which are state representatives, associate memberships are also open "to any recognized operating agency, industrial organization, or financial or development institution related to the telecommunications industry, with a legal personality, provided such membership is approved by the corresponding CITEL member state." Associate membership is said to offer "access to abundant first-hand strategic information on . . . the state of telecommunications in the countries, development and investment needs, and project initiatives (not yet announced) . . . an unparalled opportunity for joint participation in, and contribution to, the development of programs and projects in the Americas" (OAS/CITEL 1994d). The inclusion of private sector members was seen as a way of "adjusting [CITEL's] structure and working methods to meet the challenges of the future . . . [and] an essential step if CITEL is to have a significant role in telecommunications in the Americas" (OAS/CITEL 1994f).

Regional connections and concerns are also recognized in the Americas Region meetings of the International Telecommunication Union. CITEL encourages an ITU regional presence, especially in the ITU work on telecommunications development. Similarly, meetings of the Permanent Consultative Committees in Ottawa in August 1994 included formal discussion of issues which would arise in the then forthcoming 1994 ITU Plenipotentiary meetings in Kyoto, Japan (OAS/CITEL 1994e).

In the United States, telecommunications policy and planning continues to pursue primarily national objectives. However, a number of issues were addressed at the local level of government during the passage and implementation of the Telecommunications Act of 1996. Attempts were made to introduce competition in local telephone service, to support local institutions such as schools and libraries in providing access to enhanced communications services, and to introduce new wireless communications technologies. These objectives entailed a reworking of the roles and relationships with state and federal governments, and with firms providing telecommunications services. While the local/city government controls the ground, the radiomagnetic spectrum is a national resource controlled by the federal government. A number of issues have arisen in this context, including the applicability of the franchise fees charged for the use of municipal rights of way to various service providers, zoning authority over the siting of towers to provide wireless services, state and municipal taxation of communications services, the provision of telecommunications infrastructure and services by city governments, and the role of local institutions—schools, libraries, and hospitals— in efforts to promote access to advanced telecommunications services. Additionally, local governments have become actively involved in the cable industry because of their role in licensing and granting rights-of-way for communications companies.

Most importantly, the contemporary industry restructuring—a process typified by corporate mergers, by dissipating boundaries in the use of different communications, and by massive investment plans and alliances to capture a piece of the information highway—places extreme pressure on the policy and regulatory bodies in North American nation-states and in the region. The Canada-United States Free Trade Agreement and the NAFTA reflect the shrinking of nation-state policy competence, and are consistent with domestic programs of privatization and deregulation which also scale back certain roles for the state vis-a-vis the economy and social relations. Similarly, regional bodies are also responding to private industry calls for change. The permeability of the state and interstate organizations to private business purposes is also reflected in CITEL's inclusion of private sector organizations as associate members.

The contemporary structuring of relations among different levels of policy bodies in North America could be seen to reflect a quasi-feudal overlapping of forums, issues and membership. This can be juxtaposed with an ideal of a functionally organized hierarchy of sub-state, state, regional or multilateral organizations, each having their special niche of well-defined problems and specific geographic scope of responsibility. Rather than seek an exclusive geographic scope as a body to promote regional integration and cooperation, CITEL welcomes participation from any sort of sub-state, state or interstate organization, public or private, which wishes to attend. This strategy is perhaps one way to gain and maintain relevance in an era of rapid policy and industry change.

CONCLUDING COMMENTS

This chapter has argued that it is important to examine the changing forms and scopes of practices and institutions when trying to understand changes in telecommunications governance. It has provided a group of categories to ground investigation of these changes, and an historical account of how the relationships among the sub-national, national and international telecommunications governance bodies and their roles and responsibilities have been significantly redefined and remade in successive historical periods in Canada and the United States.

In the late nineteenth century, policies which protected large capital combinations, promoted privately owned infrastructure construction, and provided patent protection to private companies, were the main instruments of state intervention. The United States initially stayed out of the International Telegraph Union because it was seen as serving state interests rather than the interests of the users of communications and private industry. For the United States, the extent of participation in regional Inter-American Radio Conferences grew alongside increased participation in multilateral bodies. The International Telecommunication Union's structure and goals throughout much of the twentieth century were compatible with strong national policy bodies, rather than competing with these forms. With increased international trade and investment in telecommunications services and national regulatory restructuring in the 1990s, the geographic scope or level of public governance seen to be appropriate has again shifted, but this time towards a plurality of forums. North American policy integration is not organized exclusively in one institution, but takes place in many types of public organizations: NAFTA, CITEL, ITU, and private industry forums all coexist and interact.

What sort of relationships can be identified between the organization of economic production and the form and scope of telecommunications governance? In the late nineteenth and early twentieth centuries, national patterns of monopoly capitalism and industrial combinations were also reflected in international combinations for radio communications in South America. Imperial expansion and rivalry was manifest in corporate strategies and also in the international governance of telecommunications. The integration of national and world markets through the use of transportation and communication technologies was driven both by interstate rivalry among European powers on one hand and private sector objectives on the other, resulting in a strange mix of coordination in international telecommunications bodies and competition in private telecommunications consortia. The United States combined political efforts to protect the regional sphere of influence it claimed with a reluctance to use state power on the domestic level to control monopolies. State institutions were stronger in undertaking international coordination than in national policy intervention (see Zacher, this volume).

The relationship between the national industry organization and nation-state telecommunications governance was clearer and cleaner in the mid-twentieth century. National and regulated telecommunications monopolies in the twentieth century could depend on a guaranteed rate of return on capital, allowing for stable growth and capital accumulation in the form of fixed telecommunications investment. They generally respected technological divisions imposed between different segments of the industry by regulation. The national borders fixed equipment and service production markets, and national policy spaces were recognized by international institutions. There was a close fit between the national and international governance orders, and this relationship was stable through the growth of mass production economies until the 1970s.

In the late twentieth century, the flexible organization of production and economic globalization have contributed to pressures for a fragmentation of any clear demarcation of policy spaces for different public bodies, and greater overlapping and integration of governing responsibilities. The strengthening of patterns of flexible production is being expressed in shifting user needs and in the reorganization of telecommunications goods and services markets and ownership structures. The impacts of the economic restructuring on governance, and the efforts of states to respond to new conditions are seen not only in the shifting forms of state regulation, but also in the shifting forms of regional coordination. Current developments in trade agreements and national regulation shift the autonomy and scope of state intervention in shaping of the telecommunications industry in the public interest. The United States to a great extent (Comor 1997), but Canadian state officials also, has been actively seeking changes in international governance in order to expand and institutionalize national advantages. Other states that might have opposed these changes to the global telecommunications policy order, find their telecommunication policies, industrial sector, or technology strategies less and less appropriate to the new forms of governance and industrial and technical deployment in global telecommunications.

REFERENCES

Armstrong, Christopher, and H.V. Nelles. 1988. *Southern Exposure: Canadian Promoters in Latin America and the Caribbean 1896–1930*. Toronto: University of Toronto Press.

Babe, Robert E. 1990. *Telecommunications in Canada: Technology, Industry and Government*. Toronto: University of Toronto Press.

Beniger, James R. 1986. *The Control Revolution: Technological and Economic Origins of the Information Society*. Cambridge, MA: Harvard University Press.

Brunn, Stanley D., and Thomas R. Leinbach, eds. 1991. *Collapsing Space and Time: Geographic Aspects of Communication and Information*. London: Harper Collins.

Buzan, Barry. 1993. From International System to International Society: Structural Realism and Regime Theory Meet the English School. *International Organization*. 47 (3): 327–352.

Canada, Info Export. 1992. "NAFTA—Telecommunications", Ottawa: EAITC. (August)

Carey, James. 1988. *Communication as Culture: Essays on Media and Society*. Boston: Unwin Hyman.

Castells, Manuel. 1996. *The Rise of the Networked Society*. Oxford: Blackwell.

Codding, George A. Jr., and Anthony Rutkowski. 1982. *The International Telecommunication Union in a Changing World*. Dedham, MA: Artech House.

Codding Jr., George Arthur. 1952. *The International Telecommunication Union: An Experiment in International Cooperation*. Leiden: E.J. Brill (reprinted 1972).

Comor, Edward A. 1997. The International Implications of the United States Telecommunications Act. *Journal of Economic Issues*. 31: 549–556.

Cox, Robert W. 1987. *Production, Power, and World Order: Social Forces in the Making of History*. New York: Columbia.

Cutler, A. Claire, Virginia Haufler, and Tony Porter, eds. 1999. *Private Authority and International Affairs*. Albany: State University of New York Press.

Drache, Daniel, and Meric S. Gertler, eds. 1991. *The New Era of Global Competition: State Power and Market Policy*. Kingston: McGill-Queen's.

Drake, William J., ed. 1995. *The New Information Infrastructure: Strategies for U.S. Policy*. New York: Twentieth Century Fund.

Fischer, Markus. 1992. Feudal Europe, 800–1300: Communical Discourse and Conflictual Practices. *International Organization*. 46 (2): 427–466.

Fortner, Robert. 1979. The Canadian Search for Identity: 1846-1914—Part I Communication in an Imperial Context. *Canadian Journal of Communication*. 6 (1): 24–31.

Graham, Stephen, and Simon Marvin. 1996. *Telecommunications and the City: Electronic Spaces, Urban Places*. London: Routledge.

Harvey, David. 1990. *The Condition of Postmodernity: An Enquiry into the Origins of Cultural Change*. Cambridge, MA: Basil Blackwell.

Horwitz, Robert Britt. 1989. *The Irony of Regulatory Reform: The Deregulation of American Telecommunications*. New York: Oxford University Press.

Inter-American Radio Conference. 1937. *Final Act of the First Inter-American Radio Conference*. Havana. (November 1–December 13)

Inter-American Radio Conference. 1945. "Minutes of the Final Meeting of the III Inter-American Radio Conference Convened in Rio de Janeiro from September 3 to 25, 1945."

Inter-American Telecommunications Convention. 1945. Concluded September 27, 1945.

Kahin, Brian, and Charles Nesson, eds. 1997. *Borders in Cyberspace: Information Policy and the Global Information Infrastructure*. Cambridge: MIT Press.

Kahin, Brian, and Ernest J. Wilson III, eds. 1997. *National Information Infrastructure Initiatives: Vision and Policy Design*. Cambridge: The MIT Press.

Kellerman, A. 1993. *Telecommunications and Geography*. London: Belhaven.

Kern, Stephen. 1983. *The Culture of Time and Space 1880–1918*. Cambridge: Harvard University Press.

Kobrin, Stephen J. 1998. Back to the Future: Neomedievalism and the Postmodern Digital World Economy. *Journal of International Affairs*. 51.

Mosco, Vincent. 1988. Toward a Theory of the State and Telecommunications Policy. *Journal of Communication*. 38 (1): 107–124.

Ness, Gayl D., and Steven R. Brechin. 1988. Bridging the Gap: International Organization as Organizations. *International Organization*. 42 (2): 245–273.

Niosi, Jorge, ed. 1991. *Technology and National Competitiveness: Oligopoly, Technological Innovation, and International Competition*. Montreal: McGill-Queen's University Press.

Organization of American States. 1974. *First Inter-American Telecommunications Conference, Caraca, Venezuela, September 5–11, 1971: Final Act*. Washington, D.C.: OAS. (OAS/SER. C/VI.4.7).

Organization of American States. 1975. *Final Report, Inter- American Telecommunications Conference, November 19–27, 1975, Rio de Janeiro, Brazil*. Washington, OAS. (OEA/Ser. K/VI.8 CITEL/ 134 rev. 1).

Organization of American States. 1979. *Final Report, Third Inter-American Teelcommunications Conference, 5-9 March 1979, Buenos Aires, Argentina*. Washington, D.C.: OAS. (OEA/Ser. C/VI.4.9).

Organization of American States. 1987. *Final Report, Fifth Inter-American Telecommunications Conference, 10–14 August 1987, Lima, Peru*. Washington, D.C.: OAS. (OEA/Ser. K/VI.II.I).

Organization of American States. 1991a. "Report of the Chairman of the Permanent Executive Committee of CITEL", *CITEL* /doc.01/91. (September 23)

Organization of American States. 1991b. "CITEL Plan of Action for the Period 1991–1995", (CITEL/doc.33/add.1/rev.1). (September 25)

Organization of American States. 1994a. Inter-American Telecommunication Commission, "Regulations of the Inter-American Telecommunication Commission CITEL", (OEA/Ser.L/XVII.1). (January 1994)

Organization of American States, Inter-American Telecommunication Commission. 1994b. "Associate Members in the Work of the Permanent Consultative Committees of the Inter-American Telecommunication Commission", (PCC-III.5.95). (August)

Organization of American States, Inter-American Telecommunication Commission. 1994c. "Associate Members in the Work of the Permanent Consultative Committees of the Inter-American Telecommunication Commission", (PCC-III.5.95). (August)

Organization of American States, Inter-American Telecommunication Commission. 1994d. "Introductory Remarks by Roberto Blois, Executive Secretary of CITEL", (OEA/Ser.L/XVII.4.1). (August 22)

Organization of American States, Government of Canada. 1994e. "Preparing the Path for Progress in the Americas: Meetings of the Permanent Consultative Committees of the Inter-American Telecommunication Commission", Government Conference Centre, Ottawa, Canada. (August 22–31)

Organization of American States. 1994f. Inter-American Telecommunication Commission, "Summary of Activities in the Telecommunications Sector Developed by the

Department of Economic and Social Affairs of the General Secretariat of Organization of American States", 1994 (OEA/Ser.L/XVII.4.1, WGDEV.06/94). (August 25)

Organization of American States, Inter-American Telecommunication Commission. 1994g. "CITEL Temporary Working Group, Kyoto Plenipotentiary Conference, Issue Analysis", (OEA/Ser.L/XVII.4.1). (August 28)

Pentland, Charles. 1973. *International Theory and European Integration*. London: Faber and Faber.

Pool, Ithiel de Sola. 1990. *Technologies without Boundaries: On Telecommunications in a Global Age*. Cambridge, MA: Harvard University Press.

Ruggie, John Gerard. 1986. International Regimes, Transactions and Change—Embedded Liberalism in the Post-War World. *International Organization*. 36: 379–415.

Taylor, Paul. 1993. *International Organization in the Modern World*. London: Pinter.

Teske, Paul, ed. 1995. *American Regulatory Federalism and Telecommunications Infrastructure*. Hillsdale, NJ: Lawrence Erlbaum.

Thompson, Graham. 1990. Sandford Fleming and the Pacific Cable: The Institutional Politics of Nineteenth Century Imperial Telecommunications. *Canadian Journal of Communication*. 15 (2): 64–75.

Tribolet, Leslie Bennet. 1929. *The International Aspects of Electrical Communications in the Pacific Area*. Baltimore: Johns Hopkins

United States. 1996. *Telecommunication Act of 1996*. Public Law 104-104. (February 8)

Vogel, Steven Kent. 1996. *Freer Markets, More Rules: Regulatory Reform in Advanced Industrial Countries*. Ithaca: N.Y: Cornell University Press.

CHAPTER TEN

NEGOTIATING REGIME CHANGE: THE WEAK, THE STRONG AND THE WTO TELECOM ACCORD

J. P. SINGH

The WTO telecommunications accord signed on 15th February, 1997 formalizes the new regime in telecommunications.[1] This regime, signed by 69 countries including 40 less developed countries (LDCs), accounts for over 90 percent of the world's telecommunications revenues.[2] Historically, telecommunications sectors were controlled or operated according to domestic priorities. The new regime, effective since January 1, 1998, allows this sector to be governed by global rules underlying WTO processes. Among other things, cross-national investments in telecommunications are allowed (or hastened given that this process precedes 1997), and trade in basic and many value-added telecommunications services are governed by free trade norms, both features backed by WTO rules of transparency and Most Favored Nation.[3] Sixty-three of the 69 governments will also introduce 'regulatory disciplines' to observe the WTO rules.

An examination of the North-South negotiations included in the WTO Telecom accord is important for many reasons in the context of this volume. First, in the so-called information age, governance issues in telecommunications are themselves reflective of the emerging patterns of governance in other issue-areas. They allow us to see how businesses, states, international organizations and other transnational groups come together. Second, telecommunications infrastructures are crucial to the global political economy. Zacher (chapter 8) and McDowell (chapter 9) also underscore the importance of examining telecommunications governance.

Third, telecommunications serve as a crucial case for investigating the claim that developing countries stand to gain very little in their negotiations with the developed world in the emerging global economy, especially in high-tech issue areas.[4] If it can be shown that developing countries effected concessions in least likely issue areas such as telecommunications, then we may conjecture that they would perhaps gain even more significant concessions in such low-tech issue areas such as agriculture and textiles. Either way, the issue is more complicated than popular wisdom regarding these negotiations which presents the developing countries as having struck a Faustian bargain when they purportedly sacrificed something in issue areas such as telecommunications to get concessions in textiles and agriculture at the Uruguay Round. Fourth, if international negotiations allow developing countries to effect changes in global rule formation, then they deserve attention as we analyze the emerging global information economy and the dynamics on which it rests. Analyses of international negotiations, as Aronson (chapter 2) also shows, illustrate important components of the direction and shape of the global information economy. Taken together, the four points made above, allow us to re-examine notions of instrumental and structural power as they affect the developing world. Not only do negotiations allow developing countries to influence the structure, but also open possibilities for the exercise of instrumental power which were unavailable to them earlier.

This chapter examines two sets of questions and related answers to examine the claims noted above.

1. Do the WTO telecom rules only reflect the preferences of great powers? This chapter shows that the increased availability of alternatives and strategies allows developing countries to get their interests articulated in global rule formation.
2. Do these rules contradict domestic preferences of developing countries? It is shown here that, in most cases, the commitments made by the developing countries at the international level are consistent with their domestic agendas.

This chapter thus shows that developing countries are not so badly off in terms of the emerging global rules in telecommunications. The denouement of the story is as follows: the WTO telecommunications accord also reflects the negotiating strategies and domestic preferences of developing countries. While domestic actor involvement in shaping state preferences varies according to the degree of pluralism reflected in state's decision-making, it is not so minuscule as to rule out micro pressures completely. Although a few of the domestic pressures may themselves be international in as much as they come from international actors such as multinational corporations, they are nonetheless arbitrated by states in LDCs, often with stiff opposition from various groups, and thus they are presented as domestic here. The telecom accord also reflects preferences of all kind of user groups from the developing world including business users and urban/rural residential users.[5]

The arguments made here allow us to arbitrate among two bold claims about the developing world in the information age. One claim contends that dissemination of telecommunications will allow developing countries to leapfrog, allowing individuals and groups to empower themselves and allow them a voice in the development process.[6] In this scenario, LDCs might willingly adjust to global liberalization processes (Zacher chapter 8). The other claim portends an ominous future for developing countries. Developing country populations become pawns in the manipulations of powerful actors (rich states and multinational corporations, for example), and are disempowered from any kind of participatory rulemaking (McDowell chapter 9). In the extreme case, control over the manufacture and dissemination of information allows these powerful actors to constrain any preferences other than their own (Comor chapter 7; Kim and Hart chapter 6). The first story is subscribed to by neoliberals and, in varying degrees, by social constructivists.[7] The second story is offered by neorealists, neo-Marxists, and also by many social constructivists.[8] The versions summarized here are, of course, simplified hybrids. This chapter shows that the truth is somewhere in between.

The substantiation of the claims made in this chapter lies in examining the negotiation processes underlying the WTO telecommunication accord. These negotiation processes, which make up the nitty-gritty details of international politics, are central to examining the bold claims mentioned above, which are often asserted rather than substantiated.[9] These processes in particular help to show how outcomes can not be specified ex ante and that the way global actors interact over specific issue areas makes a difference. Negotiations matter.

Developing countries made several gains through the negotiation process and the WTO accord in general. First, by participating in the negotiations, they avoided the take it or leave it option of the earlier negotiations. Second, as later sections show, many of the developing countries had recognized that their inward-oriented investment policies did not work. Signing on to the telecom accord would allow them access to global finance and investment for their prioritized telecommunications sectors. Third, by allowing phased in commitments, developing countries could adjust their domestic policies to international ones more slowly than if all countries had to agree to the same time frame. Fourth, the accord framework was such that commitments could be made in several sub-issue areas. This feature allowed for further tailoring of commitments instead of one size fits all which would disadvantage some. Developing countries thus negotiated carefully on the sub-issue areas and the time frames most suitable to them.

The next section develops the conceptual arguments followed by three empirical sections. Empirically, a brief overview of GATT's Uruguay Round is provided first as it posits the historical context for the WTO telecommunication regulations. The February 1997 accord is in fact embedded in the General Agreement on Trade in Services (GATS) which was negotiated during the Uruguay

Round. Second, the WTO telecommunications negotiation process is described to show how developing countries were able to employ specific strategies in their favor. Finally, the domestic liberalizations of specific countries are described to show how the rules developed at the global level may not be inconsistent with domestic agendas even when they are shaped by elite preferences alone.[10]

THE ARGUMENTS

Developing countries are now favored by two circumstances in their negotiations with the developed world. First, power configurations in the world may now be more favorably disposed toward the developing world than they were earlier, giving them advantages in their bargaining. Second, the developing world is emerging from the shadows of its erstwhile inward oriented growth strategies and adjusting to neo-liberal ones, which makes their development trajectories, to a large extent, consistent with the GATT/WTO agenda in general. In terms of regime change, one can argue that the developing world in effect has already accepted the principles and norms of the neo-liberal regimes. What we are seeing at present are negotiations about the rules and decision-making procedures.[11] Each of these scenarios is described in detail below.

AVAILABILITY OF STRATEGIES AND ALTERNATIVES TO DEVELOPING WORLD

Negotiation outcomes vary from one historical context to another. Most of the analyses to date on North-South negotiations belong to one particular global context captured by the term distribution of power. Krasner (1985, 1991), for example, shows that global rules are written by the strong even in the telecommunications issue-area. Distribution of power implies a hierarchical distribution of resources and abilities simultaneously across many issue areas which almost always result in outcomes favorable to those at the top of the hierarchy. Distribution of power theories which constrain the win sets for developing countries are based on a historical conception of international political economy, in which security issues and great powers played a key role. However, Keohane and Nye (1977) stipulated the need, in a seminal work more than two decades ago, to distinguish between structures and processes, and positing particular issue-areas, as opposed to outcomes being stipulated by macro structures across all issue areas. In a negotiation context, Zartman also noted that distribution of power scenarios affix the weak in a "definitional inferiority" (Zartman 1971: IX).

As chapter 1 showed, the present world economy is characterized by two types of pluralities, multi-actor plurality and multi-issue plurality, both of which are captured by the term diffusion of power, in that the exercise of power at the global level is not constrained to one set of actors (the state) around the salient issue of security. The presence of many actors flattens the power distribution at top, allowing for several types of coalitions and interests to be reflected, while the multiplication of issues also allows for linkages, trade-offs, technocratic skills, and other bargaining strategies which were unavailable to the developing world earlier. All in all, the set of options and strategies available to developing countries is greater now than ever before. In such a diffused context, coercive strategies on the part of the strong, or confrontational strategies on the part of the weak (which characterized many of the North-South negotiations of the past), become increasingly irrelevant. As Keohane and Nye point out, power becomes less and less fungible across issue areas as we move away from a security dimension (1977). Thus, we see that actual power is often different from potential power. Keohane and Nye call for positing and examining issue-specific structures of power in this case.

Agencies of power are usually not replaced, but overlapped, by other agencies wielding the same or other forms of power.[12] Negotiations no longer take place in a context defined solely by states and their (authoritative) power maximization prerogatives, but in a world overlapped by many actors in pursuit of many goals and issues, and exercising different forms of power. In the postcolonial era, the authoritative power of dominant state actors has given way to the multiple influences of international organizations, market-oriented actors, and domestic interest groups. Strange (1991) notes that state-state negotiations of the past are now replaced by three dimensional negotiations in the international economy. The two other dimensions are: state-firm and firm-firm negotiations. This chapter adds a fourth actor: international organizations. Strengthening international organizations to provide a more neutral (less hierarchical) means of negotiations allows many developing countries to or draw upon many more options.

Part of the reason that negotiations are becoming so important is precisely because of the diffusion of information networks in the global economy. As argued in many chapters of this volume, information exchange and diffusion has allowed for group empowerment and concerted action to take place around the world. No longer are states the only well-informed actors in global politics. Well-informed strategies are allowing a host of other actors to better their alternatives and position in the global political economy.[13]

The availability of varying alternatives and strategies defines the range within which negotiations may be settled as the alternatives affect the lowest or the highest values for a negotiation that an actor may expect, depending on whether the negotiator is a seller in which case the minimum is important, while the maximum is important for a buyer. It is no accident that the concept of bargaining power is related to actors' alternatives and in essence illustrates the value of the negotiation to each actor, which in turn, depends on the actors' alternatives (Odell 1993:

19–20). Related to the concept of alternatives is also that of reservation price or the lowest or minimum value that an actor will expect. Availability of good alternatives would then raise the reservation price (White and Neale 1991). Therefore, what is termed the zone of agreement or the difference among the negotiating parties reservation prices is contingent upon each party's alternatives. Fagre and Wells note that the greater the competition among multi-national corporations for a developing country, the lower the MNC bargaining power (Fagre and Wells 1982). Grieco (1982) notes that the "emerging assertive upper tier" of developing countries has gained much from the increasing competition among developed world firms for their markets. He shows how the Indian computer industry went from a position of dependency to improving its terms as the international market structure of the computer industry enlarged to include many firms.

The next section will explore if developing countries gave in to global rules or were able to influence these rules in liberalizing their telecommunications sectors. Either way, the increasing number of alternatives and strategies is important. Some of these alternatives and strategies can be taken as given at the start of the negotiations while others arose as negotiations proceeded due to the specific deployment of many bargaining strategies by the developing countries.[14] The following bargaining strategies generally unavailable to developing countries in a state-centric, security-driven world may be noted:

A) Inclusion and Agenda-setting. There is growing recognition, especially in global multilateral fora, following developing countries' pressures, that they cannot be excluded from the negotiations or added on later as an addendum. The inclusion also results from global interdependence which accords developing countries strategies for inclusion across many issue-areas. Inclusion in the negotiation process allows the developing countries a chance to influence the agenda, in turn allowing actors to translate their potential power into results (Bennet and Sharpe 1979). During the Uruguay Round and later the WTO telecommunication negotiations, developing countries not only were included but made many changes in the agenda (including asking for a separate group for negotiating services) as a condition for their inclusion. The telecommunications negotiations almost broke down in 1996 because not enough developing countries had come on board and made offers.

B) Trade-offs/Issue-linkage. Several actors and many issues make negotiations increasingly complex, in turn increasing the risk of no agreement. However, several actors and many issues can also allow for many trade-offs and issue-linkages. Agreement can come about through either compromising on one issue for gains in another or vice versa (trade-offs), or bringing in potentially related issues to increase one's bargaining power (issue-linkage). The fact that countries were making several different types of *sub-offers* within the rubric of the WTO telecom negotiations, which themselves were part of the GATS framework allowing several types of commitments and exemptions, allowed them to tailor the agreement to their context to a large extent.

C) Coalitions. Issue specific and multilateral negotiations in particular provide developing countries with the ability to form coalitions and increase their bargaining power. In the 1960s and 1970s, the developing countries deployed coalitional tactics in a confrontational fashion, but since the 1980s, coalitions have been used for very specific purposes by them. During the WTO telecom negotiations, developing countries often found allies in the European Union countries to counter the demands made by the United States.

D) Technocratic and Legalistic Strategies. Negotiations are now increasingly biased toward persuasion. The type of persuasion most likely to effect an accommodating response from the other actor is usually one which is based on knowledge and effected by technically competent negotiating teams. Developing countries are also often able to prevail over developed countries because the former send their best negotiators to these countries while, for the latter, these countries are relatively less important in terms of staff commitments (Yoffie 1983: 22). In general, because of information diffusion, developing country experts may explore their alternatives thoroughly and base their strategies in superior technical knowledge. Negotiators in developing countries are likely to have been educated in the finest schools in the West and, apart from speaking a vocabulary which the North understands, they are able to forcefully articulate their strategies and persuade their partners across the table.[15] Finally, as international norms and rules deepen, and institutional enforcement sets in, many developing countries are employing legalistic strategies. The use of WTO dispute settlement procedures or the courts in the developed countries are examples. Braman (chapter 4) also notes how indigenous groups in the developing world have used legalistic strategies in order to get patents for their common property resources.

In summary, *the increased availability of alternatives and strategies available to developing countries in international negotiations should allow them to increasingly get their interests articulated in global rule formation* (argument one). Obviously, this argument is incorrect if LDCs have no alternative but to either accept or not accept rules framed at the global level. In other words, if negotiations do not matter.

THE NEOLIBERAL DEVELOPING WORLD

The razzle-dazzle arena of international negotiations is often merely an icing on the cake of individual country interests, the latter themselves rooted in economic conditions of particular countries. The latter need not be characterized as domestic interests, as what happens domestically in individual cases may itself be a result of global influences. However, we may say safely that negotiations carried out at the international level must find, to use Putnam's phrase, *domestic resonance* to be effected and/or implemented (1988, 450). Such resonance can develop from bottom-up in terms of interest group alignments in developing countries or top-down in terms of global interdependence which might have produced changes in the developing countries prior to international negotiations (which in turn might be aimed at producing similar changes themselves). While such top-down or bottom-up influences are hard to separate, such a separation is effected below for analytical purposes with special reference to telecommunications. What this separation shows is that by the time of the WTO negotiations, developing countries had already come a long way in making telecommunications a development priority due to domestic and international pressures. Second, this prioritization had occurred in the context of liberalization in terms of backing away from government led initiatives and in effecting links with the developed world markets.[16] All of this was rooted in the effects of erstwhile inward oriented or import substitution industrialization (ISI) strategies, the economic shortcomings of which are all too often now forgotten in many current analyses of the shift toward global liberalization.

A. Moving Beyond Import Substitution Industrialization. The slow economic gains effected through ISI among LDCs called the efficacy of the entire strategy into question by the 1980s. The low level of investment in telecommunications in developing countries did not meet the high demand for even basic telecommunications services and continued to be low despite high rates of return in telecommunications.[17] Moreover, the number of telecommunications services demanded by domestic and international users multiplied manifold. In the meantime, technological change and globalization of markets broke down the traditional *natural monopoly* argument which called for high levels of investment usually taken up by LDC states. It was becoming apparent to policy makers that, not only was telecommunications of utmost importance to development, but that the traditional ISI model which rested on the state run monopolies or Post Telegraph and Telephone was inappropriate to meet the growing demand for telecommunications services.

Three factors were crucial in the developing world in their move away from an inward-oriented strategy in the case of telecommunications and moves toward liberalizing their telecommunication sectors and, in many cases, privatizing their telecommunication monopolies. First, no matter how elaborate the central plans, capital resources remained scarce for sectors included in the plans while others had to be excluded due to lack of resources. Second, drawing upon the studies linking telecommunications to development, national planning agencies (while making telecommunications a priority) realized the inability of their resource-constrained state-run telecommunications sectors to provide enhanced investments or better services. Third, telecommunications became important as the services or tertiary sector, heavily dependent on informational exchanges, began to gain importance. Development strategies of LDCs were once preoccupied with primary and secondary sectors. However, in spite of the low importance given to services, their dynamism and productivity lies in the fact that the biggest growth rates in LDCs (as well as developed countries) in the postwar era were recorded in the services sectors. These sectors constitute areas such as banking, finance, travel, tourism, distribution networks, telecommunications, and media. Services are now estimated to account for 66 percent of the Gross Domestic Product in high income countries, 35 percent in low income economies, 52 percent in middle income economies and 63 percent in the world as a whole (The World Bank 1997, 236–237).

B) Global interdependence. It is now readily acknowledged that telecommunications is a way for developing countries to connect with the world markets and adjust to the emerging global rules governing these markets. The term globalization captures this dynamic. At an economic level, globalization, which picked up pace in the late 1970s, signifies the emergence of international markets, the associated beneficiaries, patterns of interdependence/dependence, and the sociopolitical processes sustaining these markets. Globalization implies that, given the rapid changes taking place in production, LDCs must improve their telecommunications sectors if they are to play any role in world markets. Globalization is also directly connected with interdependence, defined as "situations characterized by reciprocal effects among countries or among actors in different countries" (Keohane and Nye 1977, 18). Thus, in a world of emerging or existing tightly knit markets, telecommunications are a key to smooth information flows which are crucial for the markets to work efficiently.[18] Three major aspects of globalization (not mutually exclusive) necessitate the improvement of telecommunications sectors in LDCs. First, the coordination of the current complexity of the international division of labor, sometimes termed *Post-Fordism*, depends on rapid flows of information which makes telecommunications a necessity.[19] Second, information technologies can help to reduce costs, improve quality, and provide value-added and competitive intelligence. These intra-firm tasks, most of which require tremendous coordination, are dependent on information networks. Third, telecommunications are important for the producing firms, not just to keep in touch with fluctuations in world demand (or elasticities), but also to keep the distribution channels efficient. Gereffi (1995, 113) notes that the rise of global commodity chains "are rooted in transnational production systems, which link the economic activities of firms to technological, organizational, and institutional networks that are utilized to develop, manufacture, and market specific commodities."

The context of globalization is important for understanding the preferences of domestic and international actors in telecommunication restructurings. International actors such as GATT/WTO, ITU, International Bank for Reconstruction and Development (significant in the LDC context), transnational enterprises, and other international and large users are thus modeled as interest groups (or user groups as the case may be) which impact upon domestic political processes.[20]

Taken together the domestic and international economic conditions and interests thereof, paint the picture of a developing world which was well on its way to liberalizing its economies even before the Uruguay Round of negotiations had been concluded. The preceding section then qualifies the top-down approach toward global liberalization. Theoretically, the emphasis in this subsection is consistent with the now emphatic tradition in world politics which combines domestic and international politics.[21] It should come as no surprise that one of the first attempts in this case, namely Putnam's (1988) famous piece on two-level games referred to above, is on negotiations. A recent empirical evaluation of Putnam's work (Evans et al. 1993), *Double-Edged Diplomacy*, confirms and extends his arguments regarding the constraints faced by states vis-a-vis domestic and international actors. In terms of developing countries, the study concludes that democratic states in the developed world face more constraints than the non-democratic world. This does not mean that LDC states can act autonomously but that they might have more room to maneuver because their base of support may lie in smaller influential minorities. Conversely, we would find that the more democratic the state from the developing world, the more constrained it might feel at the negotiating table. However, *Double-Edged Diplomacy* shows that it is rare to find states in the developing world which act without taking domestic or international factors into account.

Eventually what the developing countries bring to the negotiating table will reflect not just their base of support but the economic underpinnings of this support, what Gourevitch (1986) calls the production profile. Of course, economic conditions by themselves (as outlined above) do not point us toward the constraints on the state. What is needed is, in Kindleberger's words, "the relationship between economic interests and political power" (1978, 30–31). For simplicity, this paper assumes that the negotiating positions taken by LDC states reflect their underlying bases of support but it does delve a little more deeply later to examine the state of liberalization schedules in each country (as a proxy variable for the extent to which the pro-liberalization coalitions have impacted the industry structure in each case).

We should then expect that developing countries will make strong offers for telecommunication liberalization when the domestic impetus for liberalization is strong (argu-

ment two). This argument is incorrect if developing countries make strong offers without domestic liberalization programs already in place, or make weak offers even though liberalizations are significantly under way. The argument also allows us to substantiate whether the WTO telecommunication accord is consistent or inconsistent with sectoral policies within LDCs. It also contextualizes other arguments which may be made about the seeming inconsistency of global rules with development efforts.

It may be useful here to note that domestic preferences of developing countries may not be limited to a few elite groups alone (Singh 1999). The emerging telecommunications sectors in the developing world can reflect societal (rather than elite) preferences in two ways. First, the provision of these services is important for states in constructing their legitimacy. The high densities of telephones to households in the East Asian Newly Industrializing Countries resulted directly from states' ability to maintain their legitimacy through promising a high standard of living to its population. Second, civil society may be directly involved in telecommunication policy-making. A focus on civil society and state decision-making is therefore necessary to observe the development impact of telecommunications.

Debates on telecommunications are not only a part of formal political processes like the emergence of interest groups and their coalitions or the construction of such coalitions by party systems in the country, but also that of the interplay of civil society and states. Civil society in its traditional definition conveyed the ability of society to participate in governing processes through informal networks, solving of common problems, and building a level of social trust and mutual support.[22] Now foci on civil society are increasingly showcasing the grassroots activism among any country's societal groups and their synergy and/or conflict with other governing processes such as the state. Inasmuch as these groups involve themselves in the formal or informal governing processes in these countries, they are part of the common problem solving and social trust aspects of civil society that the traditional definitions attended to. They in fact modify our views of economic restructuring being an elite process only.

Current work on civil society and development has focused on contexts under which civil society threatens the state, complements its role, or works independently of it. The relationships between state and civil society in fact "allow us to assess the extent to which state involvement facilitates developmentally effective collective action by common citizens in a diverse collection of settings around the globe (both in the Third World or in what used to be called the Second World)" (Evans 1996, 1034).[23]

The interplay between civil society and the state is directly relevant to the case of telecommunications restructurings in examining the extent to which the latter will involve and benefit the society at large. Unfortunately, telecommunications restructurings in developing countries have been too elite-driven so far to document civil society involvement effectively. That does not mean either that counter examples do not exist or that the trend is not changing. The provision of telecommunications services to the society at large in countries like South Korea and Singapore, the provision of telecommunications services in rural areas of India or Malaysia, and, in general, the construction of state legitimacy in the emerging democratic societies of the developing world provide examples of civil society-state conflicts and synergy.

The discussion assumes that participatory development, involving framing of policies at the behest of societal groups, is necessarily beneficial. Arguably, this is a normative position but, in one way or the other, LDCs have moved toward allowing societal groups a voice in the development process. This follows their (often critiqued) experience with inward-oriented development policies which were invariably accompanied by top-down central planning initiatives. Thus, the experience with moving away from ISI and accepting global liberalization must be weighed against participatory development processes in LDCs to measure the effects of telecommunications among nonelite societal groups. In as much as global telecommunication rules and LDC participatory development processes are consistent, global rules will not negate pro-development efforts in LDCs.

THE GATT/WTO NEGOTIATIONS AND INTERESTS

While the main purpose of this section is to describe the WTO telecommunication negotiations, a brief foray first into the history of previous trade rounds, particularly the Uruguay Round, helps us make sense of the developing country positions during the WTO Telecommunication negotiations. Special attention is paid to the negotiation of services during the Uruguay Round which provided the framework for WTO telecommunications negotiations.

THE URUGUAY ROUND AND BEFORE

Traditionally, the GATT rounds had been an advanced country club. The pyramidal pattern of negotiations ensured that those countries with the strongest economies dictated the initial proposals which were then multilateralized by inclusion of other parties (Winham 1986). Developing countries, since GATT formation in 1947, did benefit from those GATT rules which allowed them preferential treatment. However, they stayed at the bottom of the negotiation pyramid.

Hoekman and Kostecki characterize the period from 1986 onward as inte-grating developing countries into the GATT and marked by reciprocation (as op-posed to confrontation or demanding special treatment)[24] (Hoekman and Kostecki 1995). The integration of the developing countries into the global eco-nomic processes can be seen in the pre-negotiation phase of the Uruguay Round (Sjostedt 1994, 44–69). By the end of the Tokyo Round, it was apparent that nei-ther could the concerns of developing countries be ignored nor could all these countries be lumped under one label (concern about NIC graduation was salient in this regard). Furthermore, due to a host of economic and ideational influences, developing countries were now acceding to the liberalizing push of the global economy. In the meantime, as the economic capabilities of Western Europe and Japan increased, the United States was no longer able to call the shots. This made the pyramidal configuration of power distribution in the world flat at the top with a number of important players and their domestic and international lobbies miti-gating against extreme concentration of power to a select few or a preponderant power. The latter feature had actually made agreement on many issues difficult among developed countries in the Tokyo Round.

During the pre-negotiation phase of the Uruguay Round, developing coun-tries made their weight felt. They adopted strategies which may have been remi-niscent of prior periods but, in this case, were less revolutionary in that their purpose was to influence the agenda. They introduced their demands for bring-ing textiles and clothing in to the GATT agenda while stalling moves on the part of developed countries to bring in services, investment and intellectual property. A group of developing countries, known as Coalition of 20, led by India and Brazil played a key role in bringing recognition to these subjects and for bringing these countries effectively into the Round. It is significant that the Round itself, after several false starts, began in Punta del Este, where the agenda was not dic-tated by the developed or the dissident countries but the Swiss-Colombian text of the middle of the road group (de la Paix group) which included several developing countries and was in fact a compromise position.

The Uruguay Round illustrates several other sophisticated tactics employed by the developing countries. Their pragmatism is apparent from the fact that, even with occasional outbursts of militancy, most of them chose to work through the negotiation process to get their demands met. For the first time in the GATT process, developing countries were able to influence the agenda and bring recog-nition to themselves. They followed increasingly well-informed and technocratic strategies, in fact leading many of the negotiating sub-groups of the Round. Apart from the dissident and de la Paix coalition mentioned above, they joined or formed coalitions in specific issue areas (such as the Cairns Group on agricul-ture). In the end the trade-offs made by developing countries on the new issues came from getting concessions on issues like textiles and clothing which were important to them. Even the fact that GATT's Group on the Negotiation of Services (GNS) was technically separate from the GATT framework was a concession to developing countries. International Organizations such as the GATT/WTO itself offer avenues for developing countries to negotiate with the strength that they did not posses earlier (Conybeare, 1985).

It must also be mentioned that transnational social/NGO coalitions came about during the Uruguay Round to put direct pressures on WTO and the governments involved. Perhaps, the strongest here were agricultural and labor groups. For example, at a November 1993 protest in Bengalore, India, half a million Indian farmers were addressed by both farm and non-farm organizations from Brazil, Ethiopia, Indonesia, Korea, Malaysia, Nicaragua, the Philippines, Sri Lanka, Thailand, and Zimbabwe (Brecher and Costello 1994, 7).

NEGOTIATING SERVICES DURING THE URUGUAY ROUND

What the GNS put forth during the Uruguay Round is important in the context of this chapter. While the GNS agenda applied to many service industries (including financial services and shipping), the agreement which emerged from their deliberations, the General Agreement on Trade in Services (GATS), is particularly important in the case of telecommunications and has been called "the focal point for discussion about governance in the global information economy" (Nicolaides 1995, 269). Formally, GATS consists of 29 articles, 8 annexes and 130 schedules of commitments. The annexures cover specific sectors, including telecommunications. GATS is enforceable by the newly-created dispute settlement body of the WTO and overseen by one of three new councils established, the Council for Trade in Services.

The developing world was initially opposed to the inclusion of services in the Uruguay Round. While it realized that it could gain from concessions made in agricultural and industrial issue-areas, there was also a feeling that by giving in, it would trade away any participation in the evolving global service economy (Drake and Nicolaides 1992, 57). However, after initial hostility, cracks began to develop in the monolithic coalition that the developing world presented. The moderates, led by Colombia, began to see gains through participation in service negotiations while the hawks, like India and Brazil, held out. Eventually at Punta del Este, the developing world would agree to service negotiations but the GNS was created in deference to the opposition by LDC to the service negotiation issue. A negotiating text was produced by 1989 after preliminary negotiations designed to cover the basic framework of how to negotiate major issues. LDCs' deft diplomacy is revealed in the fact that, while they favored this 1989 text, the US and other developed countries were disappointed by it. This was a far cry from the initial misgiving about these negotiations that LDCs had. By then, the United States had offered a major concession in terms of introducing the concept of 'special treatment' which would later form the basis of allowing tailored commitments and schedules from individual countries in the GATS framework (Woodrow 1991, 336). A revised text, more agreeable to the United States and European Community, emerged in 1990; now the various sectoral negotiations could proceed. Accordingly, in May, the Working Group on Telecommunications Services was established by the GNS and initially involved 25 countries.

It is useful to examine how the three major parts of the GATS framework, adopted at the April 15, 1994 Marrakesh meeting, affect telecommunications. First, the general obligations and disciplines (GODs) provide the framework to extend GATT clauses such as most favored nation, transparency and the standard safeguards and exceptions to the service industries (a few amendments to these clauses were allowed to tailor them to services). Second, the specific commitments allow market access and national treatment in the sectors committed to by entering parties. This is a bottom-up approach, as opposed to a top-down agreement, which covers all sectors until exceptions are allowed (Jackson 1997, 309). Third, two thousand pages of "specific commitments" which pertain to progressive liberalization (market access and national treatment) schedules were tabled by countries. Like the schedules of tariffs under GATT, these commitments are considered legally binding upon member states. They pertain to eight sectoral annexes, including one on value-added or specialized telecommunications (others include those on financial services, transport, audio-visuals, and labor mobility).[25] The benefits of GATS are only allowed for signatory countries (there were 106, of which 77 are LDCs) but member states may ask for exceptions. MFN exceptions are granted for ten years to allow favorable treatment to some countries. Countries can choose the services for, and limit the amount of, market access and national treatment.

It is often assumed that developing countries acquiesced to the developed world at the Uruguay Round when it came to GATS and high-tech items. Overall, this evaluation only considers the hostility in the developing world to inclusion of such items in the Multi-lateral Trade Negotiations prior to the Round and does not take into the account the process and the outcome of the Round. First, developing countries were able to delay the start of the Uruguay Round precisely because of their opposition to the inclusion of these items. The major concession that they extracted was that services would be negotiated under a separate umbrella (Group on the Negotiation of Services) which allowed them to deviate from general GATT principles and led to the framework agreement for GATS. The sessions leading to inclusion of services in the GATT agenda and the GNS was headed by a developing country negotiator, Felipe Jaramillo, Colombia's ambassador to the GATT.

As far as the telecommunication annex of GATS goes (in which 67 governments made commitments or 56 schedules given the common offer by EU), developing countries were able to get concessions here, too. The informal coalition that came about among the developing countries and the European Union, in opposition to the push by the United States, is significant. Initially, the telecommunication annexure was to have included basic services.[26] Developing countries found coalitional partners among the Europeans who were averse to basic services being negotiated just then. Thus, the Marakesh agreement only covered value-added services. Second, the U.S. wanted to impose cost-based pricing schemes in telecommunications. Developing countries, whose cause was spearheaded by India in Geneva, would lose important revenue bases if these schemes were introduced immediately. Again, the Europeans helped the LDC cause by not agreeing. Third, important issues on satellite uplinks and downlinks (which would later almost derail the WTO telecom negotiations) were not negotiated because of LDC opposition.

WTO TELECOM NEGOTIATIONS

This section examines the role played by developing countries in the WTO Negotiations on Basic Telecommunications. A summary of the issues involved is first given, followed by details of the negotiation process with a view toward the strategies employed and concessions gained by LDCs. The domestic liberalization schedules of the developing world are then correlated with the relative strengths and weaknesses of the offers made by LDCs to ascertain the symmetry between the WTO roles and the domestic telecommunication trajectories of the developing world.

THE SCOPE OF NEGOTIATIONS

The agreement signed on February 15, 1997 in Geneva concluded nearly three years of efforts, begun in May 1994, to liberalize basic telecommunication under the auspices of NGBT (Negotiation Group on Basic Telecommunications) which was constituted in April 1994 when it was clear that the Uruguay Round would not be able to deal with basic service issues. NGBT was reconstituted as GBT (Group on Basic Telecommunications) by the Council on Services when the earlier deadline of April 1996 had to be extended because no agreement had been reached. GATT/WTO had never held sectoral talks before and thus these negotiations were "closely watched as a bellwether for the WTO's ability to conclude sectoral negotiations more broadly."[27]

Telecommunications negotiations, like their other service counterparts, involve two tricky issues. The delivery of services requires the presence of foreign operators, and, given that telecommunications services are consumed by a host of users (such as banking) with their own networks, it is necessary that these users are able to obtain fair access to telecommunications networks. The two things mean that WTO Negotiations on Basic Telecommunications involved market access (to operators and users) in terms of making these markets competitive, lifting of restrictions on foreign investment and, most important of all, the transparency of regulatory institutions. The latter issue is particularly important because historically the telecommunication industry grew up with a mixture of regulatory privileges and protections which took on country specific characteristics over time. Nicolaides (1995, 270) notes that "the GATS does not address the central obstacle to effective governance of the global information economy: the problem of regulatory fragmentation among national jurisdictions."

The WTO telecommunication service negotiations were thus about the application of MFN and national treatment to the basic telecommunication industry and the regulations covering this industry in various countries. Basic telecommunication services themselves ran the entire gamut of service provision including voice, data, telex, telegraph, fax and dedicated networks, and a host of other specialized services (like paging, teleconferencing, video transmission). Each of these services can be further subdivided into: user groups needing them or providing them (involving issues of public/non-public use, closed user groups, lease and resale of facilities); the distance they would cover (local, domestic long-distance, international); and the media used to convey these services (terrestrial, wireless, cable, satellite). Two other issues of importance involved the push by the United States toward cost-based pricing principles and the issue of the rights to uplink and downlink from satellite which were important for the United States based satellite operators (such as Motorola) to be able to provide telecommunication services globally.

In terms of this chapter, the negotiations not only involved multiple issues but also a variety of actors which would allow developing countries to define attractive alternatives. While the WTO negotiations may seem to be about state-state negotiations, two other types of actors of importance were present. The first type involved other international organizations. In the shadow was the International Telecommunication Union, with its wealth of information and expertise, which directly and indirectly influenced the talks. Initially opposed to telecommunication negotiations, the ITU, especially under its new Secretary-General Pekka Tarjanne from November 1989 onwards, moved toward the GATT/WTO position. UNCTAD was involved in the GATS process in general. In fact, UNCTAD's and ITU's slow but sure moves toward accepting some of the tenets of neoliberalism allowed many developing countries to do the same. In the case of the developing world, the role played by the World Bank in creating and guaranteeing pro-market policies is also of importance. Second, telecommunication operators, large users and, to some extent, equipment manufacturers, were involved in direct lobbying either through their home governments or directly in Geneva.

The presence of multiple actors and multiple issues would provide developing countries with attractive alternatives and several bargaining strategies. In fact, multilateral negotiations are generally recognized to be favored by the LDCs precisely for these reasons. Within the framework of the GATS's individual commitment schedules and exceptions, it meant that countries could make bottom-up offers tailored to their individual contexts and still participate in the global rule formation. In summary, the number of market segments involving user groups and operators literally allowed several hundred types of sub-commitments by countries.

THE NEGOTIATION PROCESS

The basic telecommunication agreement can be broken down into two phases. The first phase of NGBT negotiations, lasting from April 15, 1994 to April 15, 1996, resulted in no agreement, and full offers from only 11 countries (even though 48 made some kind of offer and 28 participated as observers). The second phase, lasting from April 15, 1996 to February 15, 1997, and featuring the reconstituted GBT, concluded an agreement with offers from 69 governments (or 55 commitments given that the European Union made a single offer).

The talks were off to a slow start in 1994 mostly because the European Union dragged its heels and the developing world made either very restrictive offers or none at all. The United States Trade Representative (USTR) Mickey Kantor announced in January 1996 that the U.S. would not sign unless a critical mass of countries made offers. The critical mass was assumed to be including many developing countries which were important to the negotiation process. Although the United States, Japan and the European Union accounted for nearly three-fourths of the world's total revenue, many developing countries did account for a significant portion of revenues and traffic.[28] The top ten telecommunication revenue states included Korea, Brazil, Mexico, and Argentina (in that order) followed by Hong Kong, India, South Africa, and Indonesia (in the 11th, 12th, 13th and 15th positions respectively). Hong Kong, Mexico, and Singapore ranked among top ten countries for international telephone traffic; Korea, Argentina, and India among the top for telecommunication investment; Korea, Turkey, Brazil, and India ranked among the top ten in terms of total number of telephone main lines. Furthermore, in terms of total main lines, the rate of growth in the developing world tends to be higher, 13.8 percent annual average growth rate during the 1990–95 period in the developing world as opposed to 3.5 percent in the developed world at that time. Similarly, telecommunication revenues grew at an average annual rate of 9.7 percent during the 1990–95 period in the developing world as opposed to 4.2 percent in the developed world.

By April 1996, key developing countries had either not made any offers (as in Indonesia and Malaysia) or had made very restrictive offers (like Singapore and Hong Kong) (Petrazzini 1996, 7). Furthermore, the developing countries could hide behind the shadow of the European Union, Canada and Japan whose offers vis-a-vis that of United States had been equally less forthcoming. At that time, the United States also withdrew its open market access to satellites. It was widely believed to have been at the behest of Motorola. The official reasoning was that if other countries did not open up their satellite markets then foreign competitors could come into the United States markets, while United States firms were shut out of theirs.

On April 30, 1996, the Deputy USTR Jeff Lang, who was leading the United States team, walked out of the talks. WTO then played a key role in extending the talks by another 10 months until February 1997 while leaving the accord implementation date intact for January 1, 1998. WTO made its case on the basis of domestic politics in many countries which included India and the United States with their election campaigns underway or countries such as Brazil, South Africa, and Thailand with domestic telecommunication legislation in progress (Petrazzini 1996, 8).

The talks were restarted in July 1996 and the GBT met monthly after that, including informal meetings held at the 1996 WTO Ministerial Meeting in Singapore. While issues between the United States and the European Union dominated the agenda, getting LDCs to participate was also of utmost importance. Many of them were explicitly courted. When the accord was signed on February 15, 1997, sixty-nine governments participated with 39 making improved offers over their April 1996 ones and 25 making new offers.

Aronson notes the key role played by specific U.S. agencies and many developing countries in pushing toward a close.[29] Top officials in Washington from the USTR and the FCC, and even the President himself would get involved. The November 1996 meeting in Singapore is especially significant in this regard because it involved the developing world. As noted before, one of the obstacles in April 1996 had been limited or no offers from the East Asian countries. FCC Chairman Reed Hundt and the new USTR Charlene Barshefsky called forth a special meeting in Singapore involving trade and telecommunication officials, especially from East Asia, to convince them to make credible offers. Malaysia, Indonesia, and Singapore were brought to the table with these tactics.

Many developing countries played a leadership role, in turn, convincing others to join in the process. In Peter Cowhey's words they became, "living-breathing points of reference" (interview, April 28, 1999). Singapore would become a leader in persuading other Association of South East Asian Nations, while countries like

Korea and Hong Kong led the efforts in terms of designing or accepting regulatory principles. Hong Kong's Alex Arena was instrumental in designing several regulatory features which developing countries found acceptable. In Latin America, Aronson notes, Peru was an early leader, while Brazil was quite forthcoming even though its domestic legislative battles were not settled as yet. By April 1996, Venezuela had already made a full offer to open its markets by 2000, while Mexico had come close to offering what its national laws would permit (Petrazzini 1996, 13). With Mexico, Venezuela, and Peru on board, other Latin American countries joined in.

The February 15, 1997 accord was hailed by the United States and WTO as a major victory. Ninety-five percent of world trade in telecommunications at an estimated $650 billion would fall under WTO purview beginning January 1, 1998, the date of implementation (TechWeb News: accessed February 17, 1997). Most of the signatories agreed to open up their markets one way or the other and adopt some kind of commitment to the regulatory principles (drafted by NGBT as a Reference Paper on April 24, 1996). Many analysts dubbed the accord as a victory not only for the United States, but for other developed countries, their telecom operators and large users.

Developing countries fared well. It is useful to describe their advantages in terms of the four types of strategies mentioned in the beginning of the paper: inclusion and agenda setting, trade-offs and linkages, coalitions, technocratic and legalistic strategies. (Please remember that the availability of these strategies is itself a result of multiple actors and multiple issues involved.) First, whereas earlier trade rounds including the Uruguay Round when it was being planned, often overlooked developing country claims, the WTO telecom talks made developing countries crucial in terms of their inclusion. As noted, USTR Mickey Kantor perceived many of them to be part of the critical mass needed to get an agreement. These countries had been able to set or influence the agenda during the GATS negotiations. They did the same during the telecom negotiations. In February 1997, their opposition to certain items on the agenda was most visible on accounting matters where they remained extremely reluctant to deviate from historical cost structures. (One of the results of the LDC position was that, after the talks, the FCC had to take unilateral action to correct the outflow of telecommunication revenues to many in the developing world under an historically skewed accounting scheme known as settlement rates which conferred significant advantages upon the developing world. In as much as developing countries were not forced to undertake politically unpopular measures to change their entire cost structures, this can still be construed as a concession for them. Many regulatory officials in the developing world realized that the settlements system was outdated but naturally were reluctant to suggest to policy-makers to scrap the cash cow.)

Second, trade-offs and linkages came into the picture in many ways. The whole notion of phasing in successive liberalizations in the markets and taking specific exemption for MFN and National Treatment can be seen in this regard. In this way, these countries could be signatories to the accord without undue compromise. Of special note are offers made by many Caribbean countries: Dominica, Dominican Republic, Grenada, Jamaica, and Trinidad and Tobago. Together they account for less than 0.15 percent of the world's telecommunication revenues, but all of them, with the exception of the Dominican Republic, were able to plan phasing in their market opening after 2007. It was important for the WTO (and the United States) to show numbers on board and so did not object to the delay. Third, coalitional tactics were obvious in many ways. Countries like Singapore played a key role in convincing major Asian powers. Positions taken by Venezuela, Mexico, Brazil, and Peru may have had a similar affect upon other LDCs. In an indirect way, LDCs could also hide behind coalitions in the developed world, the European Union in particular, when many contentious issues affecting them both came up. Finally, many developing countries showed considerable savvy in presenting their country schedules and in designing the rules themselves. Many of these countries are involved in domestic liberalization and regulatory exercises and this experience helped them at the international level. We now turn to examine domestic liberalizations to determine if the offers the developing world made were inconsistent with its domestic experience.

DOMESTIC LIBERALIZATIONS

The contention that the developing world compromised away its future to global multinational corporations and large users from the developed world is easily tested by comparing LDC offers with the domestic lineup of liberalization in these countries. This comparison is important for two reasons. First, as noted earlier, many LDCs made telecommunications a development priority in the 1980s for a number of reasons and adopted neoliberal policies precisely because of the resource constraints that they faced. Thus, it can be assumed that the developing world was going into these negotiations knowing the strategic importance of telecommunications. Second, the adoption of telecommunications as development priorities and the accompanying market liberalization have been attended by significant political battles.[30] Two powerful coalitions, one opposing liberalization (including domestic monopolies and their employees, government agencies, and protected businesses) and the other favoring it (large users, MNCs, and international organizations), have helped to define the particular resolution of these battles in these countries. It is hard to find a case where an LDC merely acquiesced to international demands in defining the terms of their domestic liberalization. In sum, the state of liberalization underway in each country is a good proxy variable for the trajectory taken by telecommunication coalitions of each country.

Table 10:1 correlates the strength\weakness of offers made by 31 developing countries with the state of liberalization in each of these countries. For the purposes of offers, the summary of offers available at the WTO website on the World Wide Web were used. Strong offers were taken to be those which were going to open up markets in a considerable number of market segments (including voice telephony) within two years of the 1998 implementation date and would observe the regulatory principles fully or partially. Reasonably strong offers were those seeking to adopt market opening measures in several segments within 2–4 years and including a commitment to observe regulatory principles in the future. Weak offers delayed implementation more than four years and had a weak commitment toward regulatory principles. For the purposes of domestic liberalization, the description of individual countries as presented in an important study by the World Bank carried out by Wellenius and Stern published in 1994 was used (nine of the WTO signatories are not described in this book, therefore, not included here) (Wellenius and Stern 1994). While published in 1994, future predictions are included throughout the book allowing us to make reasonable estimates of telecommunication markets in these countries in 1996–97 when they made offers at the WTO. Strong domestic liberalization is taken to mean private competition in voice (thus precluding cases which may have privatization but feature a monopoly operator) and other services and significant presence by foreign operators. Reasonably strong liberalization is taken to mean at least liberalization of markets in value-added and specialized services with some foreign entry for operators and users allowed. In some cases, such as India, liberalization in voice telephony might also be underway. Weak liberalization is taken for those countries still waiting to introduce any significant competition in any market segment of telecommunications and waiting to pass major laws changing the role of their monopoly operators.

The correlations in the table provide significant comparisons. Strong offers were made mostly by those countries which had undertaken significant liberalizations of their domestic markets. Similar correlations can be observed between reasonably strong offers and reasonably strong liberalizations, and weak offers and weak liberalizations. The deviant cases (shown in the middle row) are easily explained. All countries making strong offers but possessing reasonably strong liberalizations had actually put in place strong liberalization schedules to be effected in the future anyway. On the other hand, India was constrained by elections in 1996 and then a fragile coalition in power after June 1996 which constrained the government's hand in making even a reasonably strong offer. Indonesia was on its way to a reasonably strong liberalization program but it had not really taken off as yet. While making a weak offer, it committed itself to the possibility of allowing additional suppliers in the future. It was noted earlier that the Caribbean coun-

TABLE 10.1
COMPARISON BETWEEN THE WTO TELECOMMUNICATION
LIBERALIZATION OFFERS AND DOMESTIC TELECOMMUNICATION
LIBERALIZATION PROGRAMS OF DEVELOPING COUNTRIES

WTO Liberalization Offers

		Strong Offer	Reasonably Strong Offer	Weak Offer
D O M E S T I C L I B E R A L I Z A T I O N	*Strong Liberalization Program in Place*	Argentina Chile Dominican Rep. Korea Peru		
	Reasonably Strong Liberalization Program in Place	Colombia Malaysia Mexico Philippines Singapore Venezuela	Bolivia Brazil Hong Kong Sri Lanka	Indonesia India
	Weak Liberalization Program in Place			Antigua & Barbuda Bangladesh Belize Brunei Cote d'Ivoire Dominica Ghana Grenada Jamaica Morocco Pakistan Thailand Trinidad & Tobago

tries acted more like fillers for the numbers game involved in calling the negotiation a success, but what about large countries like Thailand, Turkey, and Pakistan? Thailand and Turkey committed themselves to review after pending national legislation while Pakistan, another country facing domestic political uncertainty, allowed only weak competition in telex and fax and delayed market access to 2004.

Domestic politics also influenced the offers in terms of state autonomy and legitimacy in developing countries. India was constrained by elections and its democratic politics in making any kind of strong offer. Brazil, on the other hand, under the strong leadership and window of opportunity given Cardoso, used its reasonably strong offer to put pressure toward passing its domestic legislation. Peruvian officials used their WTO offer to break out of a political deadlock at home on the liberalization issue. Such a move was in fact suggested to visiting Peruvian regulatory officials on a trip to the FCC in Washington, DC. (Cowhey interview April 28, 1999). Many states, in making telecommunication a development priority, are also using the sector as a way of maintaining their legitimacy. Telecommunication operators and equipment manufacturers from Singapore, Malaysia, and Korea are at the forefront of telecommunication service exports in the developing world and the governments promote these sectors as paving the way for these countries for the future. Singapore, touting itself as an intelligent island, is the most noticeable here. This may account for the leadership role or improved offers by these East Asian countries.

The oft-made argument that the domestic of the developing world is actually international (MNCs, international organizations) should be most obvious in the weakest among the weak countries and least obvious in ones with well developed telecommunication sectors. In other words, the MNCs would find it easy to bulldoze the weak countries into making strong liberalization offers. The opposite seems to be true from the correlations given in table 10.1. The weakest offers were made by the weakest countries, including countries where many MNCs already operate. On the other hand, the strong offers were made by countries like Korea, Mexico, Singapore where the telecommunications infrastructures are relatively sophisticated and nationally driven.

A few comments about civil society participation, the other side of the domestic spectrum, may also be made. Evidence of telecommunications policy framed on the behest and/or behalf of the masses continues to pour in.[31] Residential users are particularly important for two types of states: ones whose legitimacy rests on delivering high rates of economic growth (East Asian NICs) and those with relatively pluralistic domestic politics (India, Malaysia). A cursory look at the rates of growth of telecommunication services provided to rural areas would serve as the first confirmation belying an urban/elite claim. Second, as a wealth of studies on telecommunications and development show, rural and urban residential users, small business and social delivery systems use telecommunication services for a variety of purposes including such things as cutting down of business costs for small scale enterprises, market searches for farmers, improving administrative efficiency in local and provincial administrations, delivery of health care, education, and emergency services, among other things.[32]

Most developing country states now find themselves in the midst of multiple coalitional pressures, both of the elite and non-elite variety. In fact, the variation in the middle row of table 1 may be explained by the strength of pressures to liberalize (as in those making strong offers) and hedging among plural pressures (as in countries like India). With multiple coalitions, restructurings may be slow and piecemeal, but there is also a positive side to the story. Articulated coalitional demands, especially plural ones, are forms of restraints on political systems. In as much as political systems now begin to respond to wider demand pressures, they are moving away from exclusive considerations rooted in the supply-driven Post Telegraph and Telephone model. Second, these coalitions are often part of other nationwide processes and might in the long run turn out to be not so elitist at all. Third (controversial as this claim may be but consistent with this chapter's normative position), in as much as these coalitions help to make telecommunication sectors market-driven, it may be expected that telecommunication carriers will find it hard to marginalize demands to the extent that state driven PTTs did.

CONCLUSION

The arguments of this paper were explained at two levels: systemic and domestic. At the systemic level, the WTO telecom accord negotiations were examined. The paper compared the initial hostility of the developing world to telecommunication negotiations when first made part of the Uruguay Round agenda with the outcome in February 1997 when the hostile reactions were few and far between and many developing countries tabled reasonable offers. Specifically, it examined the WTO negotiation process after May 1994 when special attention was paid to specific bargaining strategies employed by developing countries to effect outcomes in their favor.

The validation of the first argument about increased alternatives to developing countries points toward two things. First, international relations are in a state of flux. The varying combinations of actors and issues allow many alternatives and strategies to developing countries to effect bargains in their favor. Second, in such a state of flux, negotiations matter. If we are to reach an understanding of how actors exercise power at the systemic level, an enquiry into the negotiation processes is essential.

Furthermore, it was found that the domestic liberalization schedules of specific countries were consistent with the offers that they made at the international level. This validates the second argument about the consistency of domestic pressures and international liberalization schedules. The WTO accord can then be seen as business as usual for developing countries. In fact, it may be argued that these countries would have continued on this trajectory even if the accord had not taken place.[33] The corollary argument about civil society participation and benefits is harder to evaluate. It requires a more in-depth examination of interest group alignments in developing countries than this chapter offers. One can say that the accord, in as much as it is consistent with domestic liberalization policies of these countries, does not completely contradict all bottom-up development initiatives in these countries. Combined with the history of this sector given at the beginning of this paper, we can not then argue that these countries would have been better off without theses international negotiations.

In the context of this volume, and more broadly the information age, this chapter makes the following contribution to our understanding of power and governance. Power in this chapter revolves around notions of power as capability, but not necessarily one in which the developing world is completely disadvantaged. When examining power within a specific structure in an issue area like telecommunications, developing world actors are not constrained from effecting things in their favor. Instrumental power then takes a new life here, as formerly weak actors exercise it. In terms of governance, a neorealist position on the weak always suffering ultimately depends on the weak having no alternatives but to suffer. In a multiperspectival world, the weak have alternatives, and governance thus becomes much more pluralistic and advocacy driven. Finally, this chapter does not delve deeply into meta-power issues because its aim is to explain a short period of negotiations between 1994 and 1997. When writing of North-South issues over the long run, meta-power issues would need to be taken into account.

Global telecommunication rules are emerging in a pragmatic international environment quite distinct from the kind of North-South hostility, which marred most negotiations until the early 1980s. International negotiations are key for deciphering the rival claims made by paradigms in international relations. This chapter builds on the important tradition in world politics, which combines domestic and international politics to explain negotiation outcomes. Milner and Keohane (1996, 6) rightly note that international economic exchanges can only be understood by a "dialogue between international political economy, heavily influenced by economic models, and comparative politics, driven these days by 'new institutionalism'." As noted earlier, one of the first attempts in this case, namely Putnam's (1988) famous piece on two-level games, is on negotiations.

Most importantly, this paper has allowed us a starting point to evaluate the development potential of information technologies for developing countries in an increasingly globalized international political economy. It does so by investigating the propositions offered above, but also by placing these propositions within the overall context of the rival claims made about these technologies as noted earlier. At the systemic level, the current scenario is different from negotiations involving the North and South in the immediate postcolonial era when the South could be silenced, ignored, or completely marginalized. But a major caveat against too much optimism regarding the emerging global rules is necessary. Participatory development processes are only now beginning to emerge in the developing world as a whole. In as much as pluralistic pressures within developing countries are hard to resolve, international negotiations and/or the implementation of global rules for developing countries are bound to be quite messy and prolonged. In other words, the potential or actual benefits of international negotiations may be constrained by the increasingly messy domestic politics of many developing countries.

NOTES

I thank Karen Litfin, Debora Spar, Virginia Walsh and Mark Zacher for their comments on earlier drafts and Peter Cowhey for sharing his firsthand knowledge of the WTO telecom negotiations in a detailed interview.

1. The definition of regimes offered by Krasner (1985, 4) is relevant here: "Regimes are principles, norms, rules, and decision-making procedures around which actor expectations converge." Regimes rules ("specific prescriptions for behavior in clearly defined areas") are discussed at length here.

2. Others joined later. As of May 1, 1999, 86 governments had signed on (*http://www.wto.org/wto/services/tel08.htm*); accessed May 1, 1999.

3. Basic services leave the content of the message, as sent originally by a user, unchanged during transmission. Valued-added services change the content or add value to enhance the information. A simple example is voicemail.

4. The description of the crucial case as being least likely follows the convention developed by Lijphart (1971). However, in as much as this case itself is made up of several observations of negotiations, the analysis below endorses the critique of the case study method of counting N=1 (instead of by the number of observations within a case) given by King, Keohane, and Verba (1994, 52).

5. Many telecommunication analyses divide societies into user groups which include: urban residential users, rural users, large users, government administrations, public/private social delivery systems, and exporters. See Singh (1999) for an analysis of LDC telecommunications based on user groups. Other user group analyses include Noam (1992) and Mansell (1993).

6. It might be argued, that if extrapolated to the issue of concern here, both Deibert (chapter 5) and Litfin (chapter 3) could be seen as supporting this claim.

7. For the neoliberal viewpoint, see Saunders et al. (1994). For the social constructivist, see Litfin (1994).

8. For the neorealist viewpoint, see Krasner (1991). For the neo-Marxist viewpoint, see various essays in Comor (1994).

9. The weak are neither absolute winners nor do the strong do what they can to make them suffer. This chapter builds upon negotiation analyses which show that weak powers can effect outcomes more favorable to themselves by following strategies such as taking advantage of the negotiation sequences and procedures (Zartman 1987), playing off one great power against another (Wriggins 1976), by following well-informed and technocratic strategies (Odell 1985), by finding loopholes in great power protectionism towards them (Yoffie 1983), by taking advantage of competitors for their resources or markets (Grieco 1982) or by possessing critical resources themselves (Knudsen 1973). However, it also diverges from these analyses by positing that developing countries may now be increasingly effecting favorable outcomes for themselves because of the current diffusion of power (see Singh, 2000a).

10. Olson (1993) argues that even autocrats might have an encompassing interest in providing public goods (in this case, telecommunications) for reasons to do with their legitimacy or to increase their revenues and taxes.

11. See Zacher in this volume for a discussion of the principles, norms, rules, and decision-making procedures in the telecommunication regime historically.

12. This point is made in different contexts and language by Ruggie (1993: 139–174) and Rosenau (1990).

13. See chapters by Aronson, Litfin, Deibert, Zacher and Rosenau.

14. Negotiations is usually taken to be a broader term referring to a complete episode of two or multiparty give and take while bargaining usually refers to specific strategies deployed by the parties during the negotiations.

15. This point is consistent with Rosenau and Fagen's (1997) and Rosenau's (chapter 11) assessment of increasingly skillful and emotionally capable technocratic elites being able to make effective foreign policy decisions and designing suitable alternatives.

16. The normative position of this chapter is consistent with neoliberal analysis.

17. Estimates of economic rates of return vary between 17 and 50 percent. The World Bank also estimated internal financial rates of return on 13 projects in telecommunications it funded as ranging between 13 and 25 percent and averaging 20 percent. The projects were able to draw internally upon 60 percent of the funds needed (Saunders et al. 1994, 15).

18. The positing of efficiency here is more than just an academic exercise. Information flows are crucial for any economic transaction to take place and, in this sense, a well

functioning telecommunications infrastructure reduces transaction costs by eliminating the barriers to information flows.

19. The international division of labor is noted by any standard text in international political economy. For its application in terms of Post-Fordism, especially in the context of telecommunications, see various articles in *Journal of Communication* (1995) especially Barrera (1995) and Mody (1995).

20. The approach I have taken here is consistent with Strange's assertion that we can capture the dynamic character of "who-gets-what" of an international economy by looking not at the surface but underneath, at the bargains on which it is based. See Strange (1982, 496).

21. See, for example, Caporaso (1997), Rosenau (1997) and Keohane and Milner (1996).

22. Such traditional definitions form the backbone of writings by Tocqueville (1835/1945), Almond and Verba (1963) and Putnam (1993, 1995).

23. This article provides a nice summary of the debates leading up to this issue. Also see Migdal, Kohli and Shue (1984) for a broad ranging discussion of involvement of societal actors in state policy-making in the developing world. They note that state-society relations are relatively ignored in development studies as compared to the dominance of institutionalist approaches.

24. Zartman (1987) notes the presence of this more pragmatic issue-focused scenario for all North-South negotiations by the 1980s.

25. GATT members could not reach agreement on basic services, therefore, the sectoral annex on telecommunications during the Uruguay Round only covered value-added services. This was, in fact, the basis of the WTO telecommunications negotiations (1994–97).

26. This paragraph borrows from Nicolaides (1995).

27. Fred C. Bergsten. Preface to Petrazzini (June 1996, viii).

28. This paragraph cites statistics from WTO www.wto.org/wto/services accessed (February 17, 1997).

29. This paragraph borrows from Aronson (1998, 16–27).

30. See Singh (2000b and 1999) for details on these battles in many LDCs.

31. See Singh (1999) for examples drawn from eight different LDCs.

32. There is a vast literature on telecommunications and development. Seminal works include Saunders et al. (1984/1994), DRI/McGraw Hill (1991), Pierce and Jecquier (1983), Hudson et al. (1979).

33. A similar argument is now made by other authors, too. See Tuthill (1996), Drake and Noam (1998), and Hufbauer and Wada (1998). Blouin (2000) disagrees.

REFERENCES

Aaron, David L. 1995. After GATT, U.S. Pushes Direct Investment. *The Wall Street Journal.* (February 2): A14.

Almond, Gabriel A., and Sidney Verba. 1963. *The Civic Culture.* Princeton: N.J.: Princeton University Press.

Aronson, Jonathan D. 1998. Telecom Agreement Tops Expectations. In Gary Clyde Hufbauer and Erika Wada. *Unfinished Business: Telecommunications After the Uruguay Round.* Washington DC: Institute for International Economics.

Aronson, Jonathan D., and Peter F. Cowhey. 1988. *When Countries Talk: International Trade in Telecommunications Services.* The American Enterprise Institute.

Belassa, Bela. 1989. *New Directions In The World Economy.* Washington Square, New York: New York University Press.

Bennet, Douglas C., and Kenneth E. Sharpe. 1979. Agenda Setting and Bargaining Power: The Mexican State versus Transnational Automobile Corporations. *World Politics.* 32.

Berrera, Eduardo. 1995. State Intervention and Telecommunications in Mexico. *Journal of Communication.* 45: 107–124.

Blouin, Chantal. 2000. The WTO Agreement on Basic Telecommunications: a Reevaluation. In *Telecommunications Policy.* 24: 135–142.

Brecher, Jeremy, and Tim Costello. 1998. *Global Village and Global Pillage: Economic Reconstruction from the Bottom Up.* Boston, MA: South End Press.

Caporaso, James A. 1997. Across the Great Divide: Integrating Comparative and International Politics. *International Studies Quarterly.* (December) 41(4): 563–591.

Chowdhary, T.H. 1989. An Indian Perspective on Sector Reform. In Bjorn Wellenius, Peter A. Stern, Timothy E. Nulty, Richard D. Stern. eds. *Restructuring and Managing the Telecommunications Sector: A World Bank Symposium.* Washington D.C.: The World Bank.

Comor, Edward. ed. 1994. *The Global Political Economy of Communication: Hegemony, Telecommunication and the Information Economy.* London: Macmillan Press Ltd.

Conybeare, John. 1985. Trade Wars: A Comparative Study of Anglo-Hanse, Franco-Italian, and Hawley-Smoot Conflicts. *World Politics.* 38.

Drake, William J., and Kalypso Nicolaides. 1992. Ideas, Interests, and Instituionalization: 'Trade in Services' and the Uruguay Round." *International Organization.* (Winter)

Drake, William J., and Eli M. Noam. 1998. Assessing the WTO Agreement on Basic Telecommunications." In Gary Clyde Hufbauer and Erika Wada, eds. *Unfinished Business: Telecommuncations After the Uruguay Round.* Washington, D.C.: Institute for International Economics.

Evans, Peter B., Harold K. Jacobson, and Robert D. Putnam, eds. 1993. *Double-Edged Diplomacy: International Bargaining and Domestic Politics*. Berkeley: University of California Press.

Evans, Peter. 1996. Introduction: Development Strategies across the Public-Private Divide. *World Development*. 24: 1033–1037.

Fagre, Nathan, and Louis T. Wells. 1982. Bargaining Power of Multinationals and Host Governments. *Journal of International Business Studies*. (Fall): 9–23.

Gereffi, Gary. 1995. Global Production Systems and Third World Development. In Barbara Stallings, ed. *Global Change, Regional Response: The New International Context of Development*. Cambridge: Cambridge University Press.

Gourevitch, Peter. 1986. *Politics in Hard Times: Comparative Responses to International Economic Crises*. Ithaca: Cornell University Press.

Grieco, Joseph M. 1982. Between Dependency and Autonomy: India's Experience with the International Computer Industry. *International Organization*. (Summer) 36: 609–632.

Hoekman, Bernard M., and Mitchell M. Kostecki. 1995. *The Political Economy of the World Trading System: From GATT to WTO*. Oxford: Oxford University Press.

Hudson, Heather E. et. al. 1979. *The Role of Telecommunications in Socioeconomic Development*. Report prepared for the International Telecommunication Union and published by Boston: Information Gatekeepers Inc.

Hufbauer, Clyde B., and Erika Wada, eds. *Unfinished Business: Telecommunications after the Uruguay Round*. Washington, D.C.: Institute for International Economics.

Jackson, John H. 1997. *The World Trading System: Law and Policy of International Economic Relations*. Cambridge, MA: The MIT Press.

Keohane, Robert, and Helen Milner, eds. 1996. *Internationalization and Domestic Politics*. Cambridge: Cambridge University Press.

Keohane, Robert, and Joseph Nye. 1977. *Power and Interdependence*. Boston: Little, Brown.

Kindleberger, Charles. 1978. *Economic Response: Comparative Studies in Trade, Finance and Growth*. Cambridge, MA: Harvard University Press.

King, Gary, Robert O. Keohane, Sidney Verba. 1994. *Designing Social Enquiry: Scientific Inference in Qualitative Research*. Princeton, NJ: Princeton University Press.

Kundsen, Olav. 1973. *The Politics of International Shipping: Conflict and Interaction in a Transnational Issue-Area 1946-1968*. Lexington, Mass: Lexington Books.

Krasner, Stephen. 1985. *Structural Conflict: The Third World Against Global Liberalism*. Berkeley: University of California Press.

———. 1991. Global Telecommunications and National Power. *International Organization*. 43: 336–366.

Lijphart, Arend. 1971. Comparative Politics and Comparative Method. *American Political Science Review*. (September) 65(3): 682–98.

Litfin, Karen. 1994. *Ozone Discourses: Science and Politics in International Environmental Negotiations*. Columbia University Press.

Mansell, Robin. 1993. *The New Telecommunications: A Political Economy of Network Evolution*. London: Sage Publications.

Migdal, Joel S., Atul Kohli, and Vivienne Shue, eds. 1994. *State Power and Social Forces: Domination and Transformation in the Third World*. Cambridge: Cambridge University Press.

Milner, Helen V., and Robert O. Keohane, eds. 1996. *Internationalization and Domestic Politics*. Cambridge: Cambridge University Press.

Mody, Asoka. 1989. Information Industries in the NICs. In Robert W. Crandall and Kenneth Flamm, eds. *Changing the Rules: Technological Change, International Competition and Regulations in Telecommunications*. Washington D.C.: The Brookings Institution.

Mody, Bella. 1995. State Consolidation through Liberalization of Telecommunications Services in India. *Journal of Communication*. 45: 70–88.

Nicolaides, Kalypso. 1995. International Trade in Information-Based Services: The Uruguay Round and Beyond. In William J. Drake, ed. *The New Information Infrastructure: Strategies for U.S. Policy*. New York: The Twentieth Century Press.

Noam, Eli. 1992. Private Networks and Public Objectives. *Aspen Quarterly*. (Winter)

Odell, John S. 1993. International Trade Bargaining Under Uncertainty: Cognitive Hypotheses and Evidence," University of Southern California, Prepared for the Project on Fairness Claims, Harmonization, and Gains from Trade. (July)

———. 1985. The Outcomes of International Trade Conflicts: The US and South Korea, 1960–1981. *International Studies Quarterly*. 29: 263–286.

Ohmae, Keniichi. 1990. *The Borderless World: Power and Strategy in the Interlinked Economy*. New York: Harper Business.

Olson, Mancur. 1993. Dictatorship, Democracy, and Development. In *American Political Science Review*. (September) 87.

Petrazzini, Ben. 1996. *Global Telecom Talks: A Trillion Dollar Deal*. Washington, DC: Institute for International Economics. (June)

Pierce, William, and Nicholas Jacquier. 1983. *Telecommunications for Development*. Geneva: International Telecommunication Union.

Putnam, Robert. 1988. Diplomacy and Domestic Politics: the Logic of Two-Level Games. *International Organization*. (Summer) 42: 427–460.

——. 1993. *Making Democracy Work: Civic Traditions in Modern Italy*. Princeton, N.J.: Princeton University Press.

——. 1995. Bowling Alone: America's Declining Social Capital. *Journal of Democracy*. 4.

Rosenau, James N., and W. Michael Fagen. 1997. A New Dynamism in World Politics: Increasingly Skillful Individuals. *International Studies Quarterly*. (December) 41.

——. 1990. *Turbulence in World Politics: A Theory of Change and Continuity*. Princeton: Princeton University Press.

——. 1997. *Along the Domestic-Foreign Frontier: Exploring Governance in a Turbulent World*. Cambridge: Cambridge University Press.

Ruggie, John Gerard. 1993. Territoriality and Beyond: Problematizing Modernity in International Relations. *International Organization*. 17.

Saunders, Robert J., Jeremy Warford, and Bjorn Wellenius. 1994. *Telecommunications and Economic Development*. Washington, DC: The World Bank.

Singh, J.P. 1999. *Leapfrogging Development?: The Political Economy of Telecommunications Restructuring*. Albany, NY: State University of New York Press.

——. 2000a. Weal Powers and Globalism: Impact of Plurality on Weak-Strong Negotiations in the International Economy. *International Negotiation*. 5: 449-484.

——. 2000b. The Institutional Environment and Effects of Telecommunication Privatization and Market Liberalization in Asia. *Telecommunications Policy*. 24: 885-906.

Sjostedt, Gunnar. 1994. Negotiating the Uruguay Round of the General Agreement on Tariffs and Trade. In I. William Zartman, *International Multilateral Negotiation: Approaches to the Management of Complexity*. San Francisco: Jossey-Bass Publishers. 44-69.

Stallings, Barbara, ed. 1995. *Global Change, Regional Response: The New International Context of Development*. Cambridge: Cambridge University Press.

Strange, Susan. 1982. Cave! Hic Dragones: A Critique of Regime Analysis. *International Organization*. 2.

——. 1991. An Eclectic Approach. In Craig N. Murphy and Roger Tooze, *The New International Political Economy*. Boulder, Colo: Lynne Rienner Publishers, 1991.

Tocqueville, Alexis de. 1835/1945. *Democracy in America*. Translated by Henry Reeve. New York: Alfred A. Knopf.

Tolan, Sandy. 1994. Against the Grain. *Los Angeles Times Magazine*. (July 10)

Tuthill, Lee. 1996. Users' Rights?: The Multilateral Rules on Access to Telecommunications. *Telecommunications Policy*. 20: 89-99.

Tyson, Laura D'Andrea. 1992. *Who's Bashing Whom? Trade Conflict in High-Technology Industries*. Washington DC: Institute for International Economics.

Tyson, Laura D'Andrea, and John Zysman, eds. 1983. *American Industry in International Competition: Government Policies and Corporate Strategies.* Ithaca, NY: Cornell University Press.

Vernon, Robert. ed. 1990. *The Promise of Privatization: A Challenge for American Foreign Policy.* Council on Foreign Relations.

Wellenius, Bjorn, and Peter A. Stern, eds. 1994. *Implementing Reforms in the Telecommunications Sector: Lessons from Experience.* Washington, D.C.: The World Bank.

White, Sally Blount, and Margaret A. Neale. 1991. Reservation Prices, Resistance Points, and BATNAs: Determining the Parameters of Acceptable Negotiated Outcomes. *Negotiation Journal.* (October): 379–389

Winham, Gilbert R. 1986, *International Trade and the Tokyo Round Negotiation.* Princeton: Princeton University Press.

Woodrow, Brian R. 1991. Tilting Towards a Trade Regime: The ITU and the Uruguay Round Services Negotiations. *Telecommunications Policy.* (August): 323–342.

The World Bank. 1997. *World Development Report 1997: The State in a Changing World.* Washington DC: The World Bank.

Wriggins, Wriggins. 1976. Up for Auction: Malta Bargains with Great Britain, 1971. In I. William Zartman, *The 50% Solution: How to Bargain Successfully with Hijackers, Strikers, Bosses, Oil Magnates, Arabs, Russians, and Other Worthy Opponents in This Modern World.* New Haven: Yale University Press.

Yoffie, David B. 1983. *Power and Protectionism: Strategies of the Newly Industrializing Countries.* New York: Columbia University Press.

Zacher, Mark with Brent A. Sutton. 1996. *Governing Global Networks: International Regimes for Transportation and Communications.* Cambridge: Cambridge University Press.

Zartman, I. William. 1971. *The Politics of Trade Negotiations between Africa and the European Economic Community: The Weak Confront the Strong.* Princeton: Princeton University Press.

Zartman, I. William, ed. 1987. *Positive Sum: Improving North-South Negotiations* (New Brunswick, NJ: Transaction Books.

CONCLUSION

CHAPTER ELEVEN

INFORMATION TECHNOLOGIES AND THE SKILLS, NETWORKS, AND STRUCTURES THAT SUSTAIN WORLD AFFAIRS

JAMES N. ROSENAU

It is more permissive than dismissive to argue that information and information technologies are essentially neutral. They do not in themselves tilt in the direction of any particular values—neither toward good or bad, nor left or right, nor open or closed systems. They are, rather, neutral, in the sense that their tilt is provided by people. It is people and their collectivities that infuse values into information. For better or worse, it is individuals and organizations that introduce information into political arenas and thereby render it good or bad. Accordingly, the neutrality of information technologies is permissive because it enables the democrat as well as the authoritarian to use information in whatever way he or she sees fit.

There is, in other words, some utility in starting with the premise that information and the technologies that generate and circulate it are neutral. It enables us to avoid deterministic modes of thought in which people are seen as being deprived of choice by the dictates of information technologies. Put more positively, the neutrality premise compels us to focus on human agency and how it does or does not make use of information technologies.

This is not to imply, of course, that consequences do not follow from the degree to which information and information technologies are available. Clearly, their availability can serve as either opportunities or constraints and, clearly, both the opportunities and constraints influence the way people conduct their political affairs, with the opportunities clarifying and facilitating the choices to be made and the constraints inhibiting and narrowing the choices. To posit the choices as facilitated or constrained by information availability is not to specify independent variables. Information and its technologies are about the contexts within which decisional alternatives are considered. They set the range within which ends and means are framed, alternatives pondered, and choices made.

As range-setting factors, information and its technologies may be forms of power, as several of the preceding chapters suggest—especially those by Braman, Comor, Deibert, and Kim and Hart—but variations in the power possessed by human agents cannot predict the outcomes—the reactions evoked by the uses of a technology's power. Reactions and outcomes derive from the audiences toward whom the power is directed and they can vary as widely as the circumstances of the audiences and their relations with those who employ the technologies. To be sure, as indicated in Aronson's chapter, different information technologies can deliver information in different forms that, in turn, can underlie variations in the way issues are structured and debated. Still, and to repeat, such differences are best regarded not as causal determinants, but as setting different ranges within which human agents make choices.

Viewed in this way, it is misleading to analyze information technologies in causal terms. Causality accounts for the choices that are made and why information is interpreted in one way rather than another. By treating information technologies as neutral, we cast them as background conditions and not as immediate stimuli to action—as second-order dynamics that influence, contextualize, facilitate, permit, or inhibit courses of action, but not as first-order dynamics that change, transform, foster, impose, or shape courses of action. The distinction between the two types of dynamics is important; it differentiates between the operation of structures and those of agents. Put more forcefully, the distinction prevents the analyst from mistaking second-order for first-order dynamics, for treating information technologies as an unseen hand that somehow gets people, groups, or communities to pursue goals and undertake actions without awareness of why they do what they do and, accordingly, without taking responsibility for their conduct.

A good illustration of the dangers of positing information technologies as first-order causal dynamics is evident in the adaptation of vertical business organizations in the 1980s to the horizontal flexibility required by the globalization of national economies. When diverse enterprises first seized upon the new technologies, they treated them as labor-saving devices and as means to control labor rather than as mechanisms for organizational adaptation. The result was an aggravation of their vertical bureaucratic rigidities. It was only after they made the necessary organizational changes in order to keep abreast of their operational environments that the information technologies "extraordinarily enhanced" the success of their enterprises (Castells 1996, 169). For all practical purposes, the restructuring of businesses away from hierarchical and toward network forms of organization preceded the considerable impact of information technologies, even as the latter then facilitated eye-catching growth on the part of the former.

In the same way, the notion of information as neutral does not ignore the convertibility of information into knowledge and, thus, into power. More accurately, information and its technologies facilitate the exercise of what has been called soft power, a concept that differentiates information from the conventional dimensions of material power such as oil production, troops in uniform, military hardware, and agricultural production. (See Nye and Owens 1996, 20–36; Rosecance 1996, 45–61; Henry and Peartree 1998; Libicki 1998, 411–28.) As clearly demonstrated during the Gulf War and the Kosovo Conflict, military capabilities today are highly dependent on advanced information technologies; the targeting of missiles and the distribution of ideas through shortwave broadcasts and the dropping of leaflets over cities exemplify the application of information to modern security strategies. Yet, despite the innumerable ways in which soft power can be used, it is nonetheless the case that information technologies on which it is based are neutral. To repeat, what counts is how officials and governments generate and employ the technologies and how publics interpret the information and knowledge that comes their way. If such were not the case, if information and information technologies were causative dynamics, then levels of knowledge in communities would be uniform and the knowledge gaps among communities would not exist.

Needless to say, as conditions with which humans must cope, information technologies are crucial dimensions of the political scene. As they change, so do the contexts in which choices are made. As new technologies are developed, so is the range of plausible choices altered. Among other things, for example, techno-logical innovations pose the question of how the range of choice is expanded by the availability of information for those who are, so to speak, informationally rich and how it is narrowed for those who are informationally poor—and, indeed, how the discrepancies between the rich and the poor configure the context within which the two perceive each other and interact.

As indicated in Singh's introductory chapter, these contextual factors have not been a preoccupation of political scientists who study world affairs, a neglect the ensuing pages seek to highlight by addressing three of the main ways in which information technologies contribute to the context within which world affairs un-fold. More specifically, the analysis explores (1) how the technologies may be al-tering the skills of individuals; (2) how they may be affecting the circumstances whereby the gap between the informationally rich and poor is undergoing trans-formation; (3) how they may be changing the conditions under which individuals and groups interact; and (4) how they may be contributing to the evolution of new global structures.

THE SKILL REVOLUTION

While the world's present population may not be more skillful than earlier gener-ations, there are good reasons to presume that the skills of today's person in the street are different than was the case for his or her predecessor. The latter may have been more skillful in building fireplaces or cathedrals, but today's citizenries are more skillful in linking themselves to world affairs, in tracing distant events through complex sequences back into their homes and pocketbooks. These changes seem so extensive as to warrant labeling them as a skill revolution, as a transformation that has three basic dimensions—an analytic dimension, an emotional dimension, and an imaginative dimension—all of which have been greatly facilitated by the recent advent of technologies that bring ideas, information, and pictures into the lives of people in ways that had not previously been possible. Global television, the VCR, the fax machine, fiber-optic cable, and the computer have all enabled people to alter their skills in such a way as to adapt more effectively to the demands of an ever more complex world.[1]

Some have argued that people tend to adapt to the information age by turning away from the realm of ideas and politics. However, quite the opposite proved to be the case in a systematic survey of Americans who make extensive use of at least four of five information technologies and were classified as connected or superconnected to the digital world:[2]

> Despite the national lament that technology undermines literacy, Connected Americans are . . . more likely to spend time reading books than any other segment of the population broken down in this survey. Seventy percent of the Connected say they spend 1 to 10 hours reading a book during a typical week; another 16 percent read for 11 to 20 hours a week. Far from being distracted by the technology, Digital Citizens appear startlingly close to the Jeffersonian ideal—they are informed, outspoken, participatory, passionate about freedom, proud of their culture, and committed to the free nation in which it has evolved (Katz 1997).

Furthermore, the dynamics of change fueling the skill revolution are likely to accelerate as increasingly e-mail and computer-literate generations of children and adolescents move into adulthood. For example, it is portentous, or at least noteworthy, that a 1999 survey of young people between 13 and 17 in the United States resulted in 63 percent who reported using a computer at home (compared to 45 percent in 1994) and 42 percent who said they have e-mail addresses (Goldberg and Connelly 1999, A1). These findings suggest that the ranks of the superconnected and the connected are likely to swell with the passage of time and the advent of new generations, thus adding to the ways in which the skill revolution is a powerful source of change in world affairs.

While the acceleration rate of the skill revolution elsewhere in the world may not match or exceed the rate in the United States, it is important to stress that the changing skills of people everywhere matter. As indicated in the ensuing analysis, the newly acquired analytic, emotional, and imaginative skills have enabled individuals to join and participate in organizations appropriate to their interests and thereby to know when, where, and how to engage in collective action. In addition, as will also be seen, the enhanced public affairs oriented skills of people have contributed to a major transformation of the global structures that govern world affairs.

BRIDGING THE INFORMATION GAP

There is little question that the benefits of the information revolution have been enjoyed by only a small proportion of the world's population and that the gap between those who are rich and poor with respect to their access to information is huge. For example, while North America and Western Europe had, respectively, 43.5 and 28.3 percent of the world information technology market in 1995, the comparable figures for Latin America on the one hand and Eastern Europe, the Middle East, and Africa on the other were 2.0 and 2.6 percent. Put even more starkly, while the number of personal computers per 1,000 people in 1995 residing in low-income and lower-middle-income economies was 1.6 and 10.0, the comparable figures for those in newly industrialized economies (NIEs) and high-income economies were 114.8 and 199.3. Consider Internet users per 1,000 people in 1996: for the former two types of economies, the number was 0.01 and 0.7 respectively, whereas the number in the latter two types of economies was 12.9 and 111.0 (World Development Report 1998/99, 63).

Notwithstanding the importance of these huge gaps between the informationally rich and poor—gaps which provide the rich with advantages and opportunities not available to the poor—such data tell only part of the story. Most notably, they do not depict the trend line, which readily allows for the assertion that not only are the informationally rich getting richer, but the informationally poor are also getting richer. The gap remains huge, but it is nonetheless the case that, in a variety of ways, the information revolution is also unfolding in the developing world and that, along several dimensions, the gap is narrowing and likely to continue to narrow in the years ahead. This shrinking of the gap stems from several sources. One is the enormous decline in the costs of information technologies, a decline that is brilliantly suggested by the fact that, for diverse reasons, "computing power per dollar invested has risen by a factor of 10,000 over the past 20 years" and that the "cost of voice transmission circuits has fallen by a factor of 10,000 over those same 20 years" (World Development Report 1998/99, 57). Another source of the narrowing gap involves the capacity of developing countries to "leapfrog the industrial countries by going straight from underdeveloped networks to fully digitized networks, bypassing the traditional analog technology that still forms the backbone of the system in most industrial countries" (World Development Report 1998/99, 57). Likewise, while most of the developing world has yet to be wired, its peoples can get a cellular phone and do not have to wait for the installation of fixed lines. It is noteworthy, for instance, that the

> Number of cellular phones per fixed line is already as high in some low- and middle-income economies as in some industrialized countries; some developing countries with low density in both traditional telephone service and cellular phones have recently invested in cellular technology at a very fast rate. . . The Philippines, a country with low telephone density (only 2.5 main lines per 100 people), has a higher ratio of mobile phone subscribers to main lines, than Japan, the United Kingdom, the United States, or several other industrial countries with densities of more than 50 main lines per 100 people (World Development Report 1998/99, 57).

Of course, the rise in the trend line in developing countries is especially noticeable among their elite and educated populations. Once the Internet was introduced into Kuwait in 1992, for example, scientists, scholars, and students came online in increasing numbers. Within six years their ranks had increased to some 45,000. Many of these are younger people who hang out in any of seven Internet cafes in Kuwait City, where they escape the heat and at the same time use the Internet for chatting, dating, or otherwise reinforcing their local culture (Wheeler 1998, 359–76). The information revolution has also reached the small villages of the Middle East: in the case of Al Karaka, Egypt, there was only one home with electricity and telephone in the 1970s, but less than two decades later all its houses had electricity and "there are also 20 telephones and more than 55 television sets . . ." (Schmidt 1993, A4).

Nor are authoritarian countries able to hold back the information revolution. China, for example, has some 1.2 million Internet accounts, many of which are shared by several users, and it would appear that the number of accounts and users grows continually (Eckholm 1998, A8). Likewise, Iran has an estimated 30,000 people with Internet accounts even as it also seeks to control the flow of information to and among them (MacFarquhar 1996, A4). Whether such controls can ever be adequately established is, however, problematic.

In sum, while there are billions of persons who do not have access to the Internet, their numbers are dwindling as more and more people and organizations everywhere are coming online. Put differently, and to recast a commonplace metaphor, to focus on those who lack access may be to see the glass as nineteen-twentieths empty, but the trend line is in the direction of it being increasingly more than one-twentieth full.

INTERACTIVE CONTEXTS

Perhaps the single most important consequence of the newer information technologies—and probably the consequence that justifies a continuing reference to the information revolution—concerns the impact on the modes through which individuals and organizations interact. Until the advent of the most recent technologies, and especially the Internet, the vast proportion of these interactions were hierarchical in nature, both within organizations and across organizations engaged in similar pursuits. The former hierarchies tended to be formally established, with ranks and positions that allowed for top-down flows of authority and policy directives, whereas the across-organization hierarchies were also marked by top-down arrangements but were more in the nature of, so to speak, pecking orders—informal but widely shared rankings of prestige, influence, and power. Both the formal and informal hierarchies, however, have been supplemented by the horizontal networks that the newer technologies permit. As a consequence of the capacities for networking facilitated by the newer information technologies, the present era is marked by a veritable explosion of organizations and associations, an explosion so vast that fully tracing and documenting it is virtually impossible. At every level of community in every part of the world new organizations are continuously being formed that are preponderantly sustained by network rather than hierarchical structures (Salamon 1994, 109–22).

Note that hierarchies are being supplemented and not replaced by networks. To stress that networks have become a central form of human organization is not to imply that hierarchies are headed for extinction. There will always be a need for hierarchy, for authority to be arrayed in such a way that decisional conflicts can be resolved and policies adopted by higher authorities when consensual agreements prove unachievable in any type of organization. The present period of dynamic transformations is likely to be one in which many hierarchies are flattened, perhaps even disrupted, but such a pattern is not the equivalent of anticipating the demise of hierarchical structures.[3]

This is not to imply that horizontal networks are new forms of organization. The networks that flow from horizontal communication have long been features of human endeavor. Such interactions have always been possible, say, by steamship and letters during most of the nineteenth century and by wireless and telephone during the first half of the twentieth century. But these earlier technologies were available only to elites. Others could not afford them. What is new today, however, is that horizontal exchanges are not only rendered virtually simultaneous by the information revolution, but their cost has been reduced to nearly nothing. As a result, horizontal networking is no longer confined to the wealthy and the powerful; instead, it is now available to any ordinary folk who have access to the Internet. Stated in terms of the new technologies, "the growth of a vast new information infrastructure including not only the Internet, but also cable, cellular, and satellite systems, etc., [has shifted] the balance . . . from one-to-many broadcast media (e.g., traditional radio and television) to many-to-many interactive media. A huge increase in global interconnectivity is resulting from the ease of entry and access in many nations, and the growing interest of so many actors in using the new infrastructure for all manner of interactions" (Ronfeldt and Acquilla 1999, 4).

The networking potential that flows from the easy availability of information technologies is perhaps especially conspicuous in the United States. Not only has Internet usage in the United States more than doubled in the last four years, (Chandrasekaran 1996, A4) but 9 percent of those in the aforementioned survey of the usage of diverse information technologies were classified as either connected or superconnected to the course of events (Katz 1997, 71). That this high-usage stratum of the public is capable of extensive networking can be readily deduced from a central finding of the survey:

> The Internet, it turns out, is not a breeding ground for disconnection, fragmentation, paranoia, and apathy. Digital Citizens [the Connected and the Superconnected] are not alienated, either from other people or from civic institutions. Nor are they ignorant of our system's inner workings, or indifferent to the social and political issues our society must confront. Instead, the online world encompasses many of the most informed and participatory citizens we have ever had or are likely to have (Katz 1997, 71).

Clearly, then, the significance of virtually free access to the Internet by ever greater numbers of people can hardly be underestimated. Already it has facilitated the formation and sustenance of networks among like-minded people who in earlier, pre-Internet times could never have converged. The result has been the aforementioned organizational explosion, a vast proliferation of associations—from environmental to human rights activists, from small groups of protesters to large social movements, from specialized interest associations to elite advocacy networks, from business alliances to interagency governmental committees, and so on across all the realms of human activity wherein goals are sought. This web-like explosion of organizations has occurred in territorial space as well as cyberspace, but the opening up of the latter has served as a major stimulus to the associational proliferation in the former. Indeed, the trend toward network forms of organization "is so strong that, projected into the future, it augurs major transformations in how societies are organized—if not societies as a whole, then at least parts of their governments, economies, and especially their civil societies" (Ronfeldt 1996). (See also Sawyer 1999, 42–6 for a skeptical view).

A stunning measure of the shift from hierarchical to network organizations facilitated by the new information technologies can be seen in innovations

adopted by the U.S. Marine Corps. In a recent exercise conducted on the California coast and called "Urban Warrior," a unit of Marines comprised of all ranks
from generals to privates launched an invasion with the lower ranked personnel
that hit the beaches all carrying handheld computers that linked them to all the
others in the unit and collectively provided all concerned with a picture of how
the battle was unfolding. In effect, they operated as a network in which rank and
hierarchy were irrelevant, an arrangement that the Marine Corps plans to apply
on a larger scale in the future (Garreau 1999, 1).

While the large extent to which the Internet underlies the trend toward networking in government, business, and military organizations cannot be over-
stated, its relevance to the world of voluntary associations and nongovernmental
organizations (NGOs) is even more profound. In effect, it has facilitated a step-
level change in what is called civil society, that domain of the private sector where
people have not had the resources, to widen their contacts and solidify their col-
laborative efforts, that have long been available to governments, corporations, and
armies. Now it is possible to inform, coordinate, and mobilize like-minded individuals in all parts of the world who have common goals to which they are willing
to devote time and energy. Equally important, NGOs and the advocacy networks
they sustain are proliferating. In 1979, for example, only one independent environmental organization was active in Indonesia, whereas by 1999 the number of
such organizations had risen to more than 2,000 linked to an environmental network based in Jakarta. Likewise, registered nonprofit organizations in the Philippines grew from 18,000 to 58,000 between 1989 and 1996; in Slovakia the figure
went from a handful in the 1980s to more than 10,000 today; and in the United
States 70 percent of the nonprofit organizations—not counting religious groups
and private foundations—filing tax returns with the Treasury Department are less
than 30 years old and a third are less than 15 years old (Bornstein 1999, B7).

Clearly, then, the proliferation of advocacy networks is altering the landscape
of world affairs and having substantial consequences for the course of events.
Whether or not a global civil society will ever evolve, it is certainly the case that
transnational networks of private citizens have become pervasive and central actors
on the global stage (See, for example Keck and Sikkink 1998.) It is not an exaggera-
tion, in other words, to note that the global stage is becoming ever more dense as a
huge variety of NGOs acquire the new technologies and thereby extend their reach
and coherence. Indeed, as indicated in the following section, it is a density that has
altered the structures through which world politics are conducted.

In sum,

> Our exploration of emergent social structures across domains of human
> activity and experience leads to an overarching conclusion: as a histori-
> cal trend, dominant functions and processes in the information age are

increasingly organized around networks. Networks constitute the new social morphology of our societies, and the diffusion of networking logic substantially modifies the operation and outcomes in processes of production experience, power, and culture. While the networking form of social organization has existed in other times and spaces, the new information technology paradigm provides the material basis for its pervasive expansion throughout the entire social structure (Castells 1996, 469).

NEW GLOBAL STRUCTURES

With people in both developed and developing countries becoming more skillful in relating to public affairs, with organizations proliferating at an eye-catching and accelerating rate, it is hardly surprising that information technologies have contributed to transformations in historic global structures. Stated most succinctly, as the global arena has become ever more dense with actors and networks, the traditional world of anarchical states has been supplemented by a second world of world politics comprised of a wide variety of nongovernmental, transnational, and subnational actors, from the multinational corporation to the ethnic minority, from the professional society to the epistemic community, from the advocacy network to the humanitarian organization, from the drug cartel to the terrorist group, from the local government to the regional association, and so on across the whole range of collective endeavor. Despite its diversity and cross-purposes, this multi-centric world is seen as having a modicum of coherence such that it co-exists with the state-centric world. In effect, global structures have undergone a bifurcation in which the two worlds are conceived as sometimes cooperating and often conflicting but at all times interacting.

Needless to say, this interaction between the worlds has been facilitated and intensified by the information technologies, thus collapsing time, deterritorializing space, and rendering traditional boundaries increasingly obsolete. Indeed, the more the technologies advance, the more have they facilitated the opening up of both governments and nongovernmental organizations to the influence of their members, to bottom-up and horizontal processes that have greatly complicated the tasks of governance on a global scale (Smith and Guarnizo 1998; see also chapters by Litfin and McDowell in this volume). For national governments, these changes—and the vast proliferation of interconnections they have fostered—have confounded the traditional practices of diplomacy and the long-standing premises of national security, thereby necessitating a rethinking of how to pursue goals in relation to the demands of both other states and the innumerable collectivities in the multi-centric world (See Solomon, Wriston and Schultz 1997; and Center for Strategic and International Affairs 1998). For the latter the increased connectivity has provided opportunities as well as challenges as they seek to network and build coalitions with like-minded actors and contest the coalitions that stand in the way of their goals.

In short, the bifurcation of global structures has led to a vast decentralization of authority in which global governance becomes less state-centric and more the sum of crazy-quilt patterns among unalike, dispersed, overlapping, and contradictory collectivities seeking to maintain their coherence and advance their goals. More than that, the interconnection of these patterns "is likely to deepen and become the defining characteristic of the 21st Century. The information revolution is what makes this possible; it provides the capability and opportunity to circuitize the globe in ways and to degrees that have never been seen before. This is likely to be a messy, complicated process, rife with ambivalent, contradictory, and paradoxical effects" (Ronfeldt and Arquilla 1999, 19-20).

The information revolution may be neutral in the sense that it permits the application of diverse and competing values, but clearly it underlies extensive consequences in every realm of global affairs. Since there is no end in sight to the development of new information technologies, clearly the full ramifications of their impact are yet to be experienced as people and their collectivities seek to keep abreast of the complexities of the dynamic transformations that are altering the human condition.

NOTES

1. An extensive discussion of the skill revolution can be found in Rosenau (1990, Chapter 13; 1997, Chapter 14-15). Newly generated empirical materials that affirm the hypothesis that skills in the realm of public affairs have advanced in recent decades are provided in Rosenau and Fagen (December 1997, 655-86).

2. The Superconnected were those in the survey of 1,444 randomly selected Americans who exchange e-mail at least three days a week and use a laptop, a cell phone, a beeper, and a home computer, whereas the Connected were those who exchange e-mail three days a week and use three of the four other technologies.

3. For an analysis that stresses the limits of networks and the necessity of hierarchies, see Fukuyama (1996).

REFERENCES

Bornstein, David. 1999. A Force Now in the World, Citizens Flex Social Muscle. *New York Times.* (July 10)

Castells, Manuel. 1996. *The Information Age: The Rise of the Network Society* Vol. I. Oxford: Blackwell Publishers.

Center for Strategic & International Affairs. 1998. *Reinventing Diplomacy in the Information Age*. Washington, D.C.

Chandrasekaran, Rajiv. 1996. Politics Finding a Home on the 'Net'. *Washington Post*. (November 22)

Eckholm, Erik. 1998. A Trial Will Test China's Grip on the Internet. *New York Times*. (November 16): A8.

Fukuyama, Francis. 1996. Social Networks and Digital Networks. Paper presented at the *Workshop on the Future of the Internet*, Palo Alto, California. (May 6)

Garreau, Joel. 1999. Point Men for a Revolution: Can the Marines Survive a Shift from Hierarchies to Networks? *Washington Post*. (March 6)

Goldberg, Carey, and Marjorie Connelly. 1999. Fear and Violence Have Declined Among Teen-Agers, Poll Shows. In *New York Times*. A1. (October 20)

Henry, Ryan, and Edward C. Peartree, eds. 1998. *The Information Revolution and International Security*. Washington, D.C.: CSIS Press.

Katz, Jon. 1997. The Digital Citizen. *Wired*. (December) 71.

Keck, Margaret E., and Kathryn Sikkink. 1998. *Activists Beyond Borders: Advocacy Networks in International Politics*. Ithaca: Cornell University Press.

Libicki, Martin C. 1998. Information War, Information Peace. *Journal of International Affairs*. 51: 411–28.

MacFarquhar, Neil. 1996. With Mixed Feelings, Iran Tiptoes to the Internet. *New York Times*. (October 8) A4.

Nye, Jr., Joseph S., and William A. Owens. 1996. America's Information Edge. *Foreign Affairs*. 75: 20–36.

Ronfeldt, David. 1996. *Tribes, Institutions, Markets, Networks: A Framework About Societal Evolution*. Santa Monica: RAND.

Ronfeldt, David, and John Arquilla. 1999. "What If There Is a Revolution in Diplomatic Affairs?" Paper presented at the *Annual Meeting of the International Studies Association*, Washington, D.C. (February 17)

Rosecrance, Richard. 1996. The Rise of the Virtual State. *Foreign Affairs*. 75: 45–61.

Rosenau, James N. 1990. *Turbulence in World Politics: A Theory of Change and Continuity*. Princeton: Princeton University Press.

——. 1997. *Along the Domestic-Foreign Frontier: Exploring Governance in a Turbulent World*. Cambridge: Cambridge University Press.

———. 1999. The Skill Revolution and Restless Publics in Globalized Space. In Michel Giraud, ed., *Individualism and World Politics*. New York: St. Martin's Press.

Rosenau, James N., and Michael W. Fagen. 1997. Increasingly Skillful Citizens: A New Dynamism in World Politics? *International Studies Quarterly*. 41:655-86.

Salamon, Lester M. 1994. The Global Associational Revolution: The Rise of the Third Sector on the World Scene. *Foreign Affairs*. (July/August)

Sawyer, Deborah C. 1999. The Pied Piper Goes Electronic. *The Futurist*. (February)

Schmidt, William E. 1993. The Villages of Egypt Relish the Fruits of Peace. *New York Times*. (September 24): A4.

Smith, Michael Peter, and Luis Eduardo Guarnizo, eds. 1998. *Transnationalism from Below*. New Brunsick, NJ: Transaction Publishers.

Solomon, Richard H., Walter B. Wriston, and George P. Schultz. 1997. Keynote Addresses from the *Virtual Diplomacy Conference*. Washington, D.C.: U.S. Institute of Peace.

Wheeler, Deborah L. 1998. Global Culture or Culture Clash: New Information Technologies in the Islamic World—A View from Kuwait. *Communications Research*. 25: 359-76.

World Development Report. 1998/99. *Knowledge for Development*. 63. New York: Oxford University Press for the World Bank.

LIST OF CONTRIBUTORS

Jonathan Aronson
Joanthan Aronson is Professor and Director of School of International Relations and Professor of Communication at the Annenberg School for Communication at the University of Southern California. He is the author of numerous books and articles including *When Countries Talk: International Trade in Telecommunications Services* (American Enterprise Institute) and *Managing the World Economy: The Consequences of Corporate Alliances* (Council on Foreign Relations) (both co-authored with Peter Cowhey).

Sandra Braman
Sandra Braman is Reese Phifer Chair of Telecommunication and Film at the University of Alabama. Over three dozen journal articles and book chapters have explored changes in the nature of the nation-state, the economy, and the nature of the law itself as a foundation for analysis of policy-making and policies in an increasingly privatized and necessarily global environment. Braman edited a special issue of the *Journal of Communication* on the use of telecommunications policy in the exercise of power by the nation-state, co-edited *Globalization, Communication, and Transnational Civil Society*, and served as book review editor of the *Journal of Communication*. Current projects include two books: one on information policy analysis for a "post-law" world and one on biotechnology.

Edward Comor
Edward Comor teaches International Communication and is a faculty member of the School of International Service, American University in Washington, D.C. He is the author of *Communication, Commerce And Power* (Macmillan and St. Martin's 1998) and is the editor of *The Global Political Economy of Communication* (Macmillan and St. Martin's, 1994). Dr. Comor has published articles in various journals including *Global Governance* and *Journal of Economic Issues* on subjects related to the political economy of communication and culture developments.

Ronald J. Deibert

Ronald J. Deibert is assistant professor of political science at the University of Toronto, specializing in technology, media, and world politics. He is the author of *Parchment, Printing, and Hypermedia* (New York: Columbia University Press 1997) and numerous articles and book chapters on the communication technologies and world politics. He is currently completing a book manuscript on security and the Internet. He received his Ph.D. in political science from the University of British Columbia, Canada, in 1995.

Jeffrey A. Hart

Jeffrey A. Hart is Professor of Political Science at Indiana University, Bloomington, where he has taught international politics and international political economy since 1981. He also has taught at Princeton University, and was a professional staff member of the President's Commission for a National Agenda for the Eighties from 1980 to 1981. Hart worked at the Office of Technology Assessment of the U.S. Congress in 1985–86 as an internal contractor and helped to write their report, International Competition in Services (1987). He was visiting scholar at the Berkeley Roundtable on the International Economy, 1987–89. His publications include: *The New International Economic Order* (1983); *Interdependence in the Post Multilateral Era* (1985); *Rival Capitalists* (1992); *The Politics of International Economic Relations, 5th edition* (1997) with Joan Spero; *Globalization and Governance* (1999) (edited with Aseem Prakash); and scholarly articles in *World Politics, International Organization,* the *British Journal of Political Science, New Political Economy,* and the *Journal of Conflict Resolution.*

Sangbae Kim

Sangbae Kim received his doctorate from the department of political science, Indiana University, Bloomington. He got his B.A. and M.A. in International Relations from Seoul National University, South Korea. His research interests are in the global politics of information and technology. His dissertation was entitled "Wintelism vs. Japan: Technology, Power and Governance in the Global Computer Industry".

Karen T. Litfin

Karen T. Litfin is Associate Professor of Political Science at the University of Washington. She is author of *Ozone Discourses: Science and Politics in Global Environmental Cooperation* (Columbia University Press 1994) and editor of *The Greening of Sovereignty in World Politics* (MIT Press 1998). Her research interests include international relations theory, global environmental ethics, and feminist approaches to world politics.

Stephen D. McDowell

Stephen D. McDowell teaches in the Department of Communication at Florida State University in Tallahassee. His research includes work on telecommunications policy and new communication technologies. He has held fellowships with the Strategic Policy Planning Division of the Canadian Federal Department of Communications in Ottawa (1987–1989), the Shastri Indo-Canadian Institute (1989–1990), and a Congressional Fellowship in Washington D.C. (1994–1995). The book, *Globalization, Liberalization and Policy Change: A Political Economy of India's Communications Sector* (New York: St. Martin's) appeared in 1997.

James N. Rosenau

James N. Rosenau, University Professor of International Affairs, holds a distinguished rank that is reserved for the few scholar-teachers whose recognition in the academic community transcends the usual disciplinary boundaries. Dr. Rosenau is a renowned international political theorist with a record of publication and professional service that is acknowledged worldwide. His scholarship has focused on the dynamics of change in world politics and the overlap of domestic and foreign affairs, resulting in more than 35 books and 150 articles. His most recent publications include: *Along the Domestic-Foreign Frontier: Exploring Governance in a Turbulent World* (1997), *Thinking Theory Thoroughly: Coherent Approaches to an Incoherent World* (1995), *Global Voices* (1993), *Governance Without Government* (1991), and *Turbulence in World Politics: A Theory of Change and Continuity* (1990).

J. P. Singh

J. P. Singh is Assistant Professor in the Communication, Culture and Technology Program at Georgetown University. He is the author of *Leapfrogging Development? The Political Economy of Telecommunications Restructuring* (State University of New York Press 1999). He is currently working on another book project titled *Communications and Diplomacy: Negotiating the Global Information Economy* (Macmillan/St. Martin's forthcoming). Research interests include economic history, international development, technology, and gender issues.

Mark W. Zacher

Mark W. Zacher is professor of political science at the University of British Columbia. He was director of the Institute of International Relations from 1971 to 1991. He has written extensively on the politics of international organizations and regimes. He is the author of *Governing Global Networks: International Regimes for Transportation and Communications* (Cambridge University Press 1996). He is a member of the editorial boards of *International Organization* and *International Studies Quarterly*.

INDEX

3Com, 157
Ackerman, Frank, 171, 183n. 3
actual power, 243
Adobe, 157
Adorno, Theodor, 17, 29n. 23
advertising. *See also* branding, marketing; con-
    sumerism and, 175; mailing lists, 126; market
    power shift and, 157; web based, 44, 175
advocacy. *See also* NGOs; anti-consumerist,
    184n. 11; capitalist, 133–134; networks, 3,
    19, 29n. 26;NGOs and satellites, 72–84; pro-
    liferation of networks, 283; promoting tech-
    nologies, 20; technological, 24–25
aeronautical communications, 193
Afghanistan, 119. *See also* Taliban
Africa, 279
agenda-setting (in negotiations), 244, 251, 258
Agre, Philip, 126
agriculture; biotechnology and defense,
    101–104; biotechnology and, 104–107;
    biotechnology and trade, 95–100; Cairns
    Group, 252; genetically modified organisms,
    103; grain purchasing practices and capital-
    ism, 171; information intensive, 105, 107,
    108; markets and telecommunications, 263;
    satellites and, 66, 79, 85; technological inno-
    vations, 92; transnational coalitions, 252
AIDS, 99
Algeria, 78
Alic, John, 159
Almomd, Gabriel, 267n. 22
alternatives in negotiations, 243–245, 263–264
Amazon.com, 53, 183n. 5
AMD (Advanced Micro Devices), 149
analog technology; effects on industry, 5;
    leapfrog, 279–280
anarchy, 217–218, 284
Anderson, Benedict; identity formation,
    118–119; in relation to meta-power (*See also*
    meta-power), 15–16; technological impact,
    16–17
Andrews, Paul, 164n. 3
Angelides, Marios, 129
anthrax, 102. *See also* biological warfare

Antigua and Barbuda, 261
antitrust, 159, 162–163
AOL, 45, 57
Apple, 145, 150, 151
Archibugi, Daniele; structural analysis, 28n. 14
Argentina, 224, 226, 256, 261
Aronson, Jonathan, "field of dreams" approach,
    28n. 10; competitive networks, 8; cross-
    referenced, 2, 3, 5, 23, 24, 77, 164n. 5, 164n. 6,
    170, 219, 240, 266n. 13, 276; importance of in-
    formation technology, 7; networks and politics,
    27n. 2; telecom liberalization, 50, 199, 206;
    WTO telecom negotiations, 257–258, 267n. 29
Arquila, David; enhancement, 8; interactive
    media, 282; network effects, 285; networked
    security, 3, 122
artificial intelligence, 93
Asia. *See* East Asia
Aspry, William, 117
Association for Progressive Communications, 26
associationalism, 282
AT&T; Bell company, 224; breakup, 60n.1,
    206; bypassing national monopolies, 200;
    monopoly power, 7, 50, 199; preventing
    "open access," 57
ATMs, 42, 55
audio-visual services, 253
Australia, 222
auto industry, 143

Babe, Robert, 214–215, 224
Baby Bells, 200
Bahamas, 226
Baker, John, 86n. 3
Bandwidth. *See also* infrastructure; encryption
    and, 140n. 57; high and narrow, 3
Bangladesh, 261
banks; controlling money, 41–42; financial
    crises, 60n. 3; flesh-and-blood tellers, 177;
    GATS and, 252; information intensive, 200,
    247, 255; security concerns, 129, 139n. 47;
    transaction networks, 129, 133
Bar, Francois, 57, 117, 195
Baran, Nicholas, 175

293

SUNY series in Global Politics
James N. Rosenau, Editor

Beyond Boundaries? Disciplines, Paradigms, and Theoretical Integration in International Studies—Rudra Sil and Eileen M. Doherty (eds.)

Why Movements Matter: The West German Peace Movement and U. S. Arms Control Policy—Steve Breyman

International Relations—Still an American Social Science? Toward Diversity in International Thought—Robert M. A. Crawford and Darryl S. L. Jarvis (eds.)

Which Lessons Matter? American Foreign Policy Decision Making in the Middle East, 1979–1987—Christopher Hemmer (ed.)

Hierarchy Amidst Anarchy: Transaction Costs and Institutional Choice—Katja Weber

Counter-Hegemony and Foreign Policy: The Dialectics of Marginalized and Global Forces in Jamaica—Randolph B. Persaud

Global Limits: Immanuel Kant, International Relations, and Critique of World Politics—Mark F. N. Franke

Power and Ideas: North-South Politics of Intellectual Property and Antitrust—Susan K. Sell

Agency and Ethics: The Politics of Military Intervention—Anthony F. Lang, Jr.

Theories of International Cooperation and the Primacy of Anarchy: Explaining U. S. International Monetary Policy-making After Bretton Woods—Jennifer Sterling-Folker

Information Technologies and Global Politics: The Changing Scope of Power and Governance—James N. Rosenau and J. P. Singh (eds.)

Technology, Democracy, and Development: International Conflict and Cooperation in the Information Age—Juliann Emmons Allison (ed.)

Debating the Global Financial Architecture—Leslie Elliot Armijo

Systems of Violence: The Political Economy of War and Peace in Colombia—Nazih Richani

Political Space: Frontiers of Change and Governance in a Globalizing World—Yale Ferguson and R. J. Barry Jones (eds.)